SACRAMENTO PUBLIC LIBRARY
D0884692

In Praise of Copying

IN PRAISE OF COPYING

Marcus Boon

Harvard University Press
Cambridge, Massachusetts / London, England / 2010

Printed in the United States of America

Library of Congress Cataloging-in-Publication Data

Boon, Marcus.
In praise of copying / Marcus Boon.
p. cm.
Includes bibliographical references and index.
ISBN 978-0-674-04783-9 (alk. paper)
1. Copying. 2. Philosophical anthropology. 3. Mahayana Buddhism—Doctrines. I. Title.
BD225.B66 2010
153—dc22 2010005047

Contents

There are many that I know and they know it. They are all of them repeating and I hear it. I love it and I tell it. I love it and now I will write it. This is now a history of my love of it. I hear it and I love it and I write it. They repeat it. They live it and I see it and I hear it. They live it and I hear it and I see it and I love it and now and always I will write it. There are many kinds of men and women and I know it. They repeat it and I hear it and I love it. This is now a history of the way they do it. This is now a history of the way I love it.

Gertrude Stein, *The Making of Americans,* 1934

Introduction

The pilgrims line up for miles and miles around the mountain. They have come here from all over the world to this fabled place, at the edge of a swamp. Individuals, couples, families. Some first came when they were children. Now they bring their children. Or their children's children. Some look anxious, others bored; others are full of gleeful anticipation. My palms are sweating—I don't exactly know why. The line moves slowly and we enter the darkness of a tunnel. Inside I can hear the whirring of machines. As with anything that one is scared of, there is a nervous, almost erotic energy that buzzes through my body. But I feel foolish too, surrounded by children, ordinary folks, who hide their own fear so well, or else mastered it long ago. Finally the tunnel opens up. A black night sky, the whirling of galaxies, costumed security guards. A bullet-shaped car pulls up. Now it's our turn to step up, step in, ride the rollercoaster at Disney World's Space Mountain.

I first came here in 2005 at the suggestion of my Tibetan Buddhist teacher Khenpo Tsültrim Gyamtso Rinpoche, a seventy-year-old Tibetan lama who lives a nomadic existence traveling from Buddhist center to Buddhist center around the world. One afternoon during a teaching in Toronto, a member of the audience, perhaps exasperated by the elevated tone of Rinpoche's philosophical talk, asked how we could really experience the luminous emptiness of all phenomena and of the mind—described in the Buddhist sutras that Rinpoche quotes from. "Go to scary movies, amusement park rides," Rinpoche replied. "And when you're frightened, meditate by saying, 'This is a dream' or 'I died and this is a *bardo*.' Go to Euro Disney, Space Station 2, and you'll be thrown into nonconceptual states!" We laughed, the way Western students do, enjoying the supposed irony of a Buddhist master talking about state-of-the-art amusement park rides and bringing Disney and Tibet together. But Rinpoche continued: "This is a great way to practice Mahāmudrā—which is meditation on whatever's happening in your mind. Mahāmudrā is a very vivid meditation because you look directly at your own mind and relax—that's the supreme meditation. If you meditate in this way, suffering won't be unbearable. When you're up in the dark, flipping around, you don't have much time to think of anything. If you practice like this, you'll be able to do it in a moment of great fear. In the modern world it is impossible to avoid dangerous, frightening activity, but if we embrace fear and difficulty and cultivate the meditation of looking directly into its essence, and relax into it, then it's not difficult. And if you train now, when you face difficulty, such as death, you'll be able to meditate."

So a few months later I got on a plane and headed south. I was prepared for the fakeness, of course, but not so much for the feeling that, in fact, Disney World *is* like Tibet. Disney World's various attractions, like the most famous Tibetan monasteries, cost you a lot of money to visit as a tourist, and are patrolled by undercover security forces making sure that nothing gets out of hand. In *Simulations,* Jean Baudrillard argues that the Disney theme parks are a fine example of what he calls "modeling"—which is to say the production of designed spaces which can be implemented at various places, rather than organically existing in a single place, the way a particular church or town does. Such modeled spaces are obviously "constructions," yet they occupy space in the same way that something "real" or "original" does. Prefabricated suburban condo villages and shopping malls are a good example of this. Disney, which has created replicas of its theme parks around the world, is another. But the Tibetan monasteries are too. Samye, the oldest monastery in Tibet, for example, was built as a replica of an Indian Buddhist temple called Otantapuri. There are other replicas of this mandala-like structure to be found in other parts of Tibet. Mandalas are patterns, mental frameworks. Just as a Disney theme park is an iteration of a framework, albeit one with a not particularly stellar meaning, the Tibetan monasteries are also "hard copies" of a mental framework. This principle—that of the model—is apparently one that works exceedingly well.

As I rode the rides at Disney World, attempting to experience the fact that Space Mountain and the mythical Mount Meru of Buddhist scriptures, hegemonic oppressive late capitalism in all its cheesy negativity and the highest meditation practices of the Tibetan Kagyu lineage, are, to use a Buddhist formula, "of one taste," I found myself thinking about an apparently very different project that I was working on, relating to imitation in contemporary culture. Wasn't part of the point of this meditation that we are always in some kind of mimetic framework, even in the act of dying, being tossed in the air, or at home asleep? And that one could investigate such a framework? But suppose copying is what makes us human—what then? More than that, what if copying, rather than being an aberration or a mistake or a crime, is a fundamental condition or requirement for anything, human or not, to exist at all? If such is the case—and this is what I will argue in the pages that follow—then the activities known as "copying," the objects known as "copies," and those who find themselves making these copies would all need to be revalued. But—is there anything that does not involve "copying"? And if *that* is the case, why exactly does copying another person's actions or works make us so uncomfortable? Furthermore, having recognized copying for what it is—what kind of freedom do we have to transform the imposed mimetic structures that frame us, internally and externally, individually and as societies? For me, a Buddhist meditation on copying implies not assimilation to hegemonic structures, but the insight to see them for what they are and then to change them.

This book grew out of the observation that copying is pervasive in contemporary culture, yet at the same time subject to laws, restrictions, and attitudes that suggest that it is wrong, and shouldn't be happening. On the one hand, many of the most visible aspects of contemporary culture—the art of Takashi Murakami or Elizabeth Peyton, electronic music ranging from hip-hop and techno to dubstep and mashups, BitTorrent and other digital networks of distribution, software tools like Google Earth or Photoshop, social networking sites like Facebook and Twitter, movies like *Borat* or *Slumdog Millionaire* (all no doubt hopelessly out of date by the time you read this)—rely explicitly on something we call "copying." Indeed, many of the most vibrant aspects of contemporary culture indicate an obsession with the act of copying and the production of copies, and it seems that we find real insight into what human beings and the universe are like through thinking about how and what we copy. On the other hand, every time we install a new piece of software, listen to music, or watch a movie, we encounter the world of copyright and intellectual-property law, and the set of restrictions that have been placed around our access to and use of objects, processes, and ideas produced by the act of copying. Simultaneously, as our ability to make copies expands at both the macro (geophysics and the manipulation of global weather systems) and micro (nanotechnology and the fabrication and replication of matter from the atom up) levels, these same laws are used by corporations to appropriate, copy, and sell increasingly large parts of what was once the "public domain."

I have been teaching a course on copying at York University in Toronto for the past two years. The university is a place that is truly saturated with copies and copying. In large lecture courses, the students come to class dressed up in chaotic but well-defined subcultural fashions, which they can read almost instantaneously on each other (and on me). They move through a maze of corporate branding which controls everything from drinking water to the bathroom walls. They are encouraged to learn through the act of repeating information, quoting, appending citations, in the traditional academic way; but with access to the Internet, to computers that can copy, replicate, and multiply text at extraordinary speed, they are also exhorted not to imitate too much, not to plagiarize, and to always acknowledge sources. They are ordered not to copy—but they are equally aware that they will be punished if they do not imitate the teacher enough!

Students today live in a culture of downloads, filesharing, networks in which information, data, music, images can be exchanged almost instantaneously. When I talked to them about such things, I was surprised to find how ambivalent most of them felt about it all. I expected them to be proud or excited by the things they spend most of their lives doing, able to celebrate the value of the incredible tools at their disposal. Like our broader society, though, they seem thoroughly confused or conflicted about this state of affairs—and about talking about it with a professor whose role, in their eyes, is to maintain the law, which says that it is all wrong (except when it's right). Thus, they live in a constant state of vague, unarticulated guilt or duplicity, filesharing, downloading MP3s, imitating styles, yet also grudgingly accepting the arguments against what they are doing, which are based on important but unexamined concepts like property, ownership, originality, authenticity—concepts which have been given very particular meanings by states and corporations at the beginning of the twenty-first century.

I know that people reading this book will expect to find here an ethics of copying—but from the outset, I would like to call such a desire into question. Can we really identify an area of human activity outside copying which would make it possible for us to choose or decide whether to copy or not? I will argue that there is no such area, that we are always entangled in the dynamics of mimesis, and I write "in praise of copying" as an affirmation of copying rather than as an ethics. The word "copyright" (nearly 3.8 billion hits on Google) itself sounds a little desperate, as though one had to actually suture the words "copy" and "right" together in order for them to associate consistently. Just to put that number in perspective, "freedom" gets only 315 million Google hits and "truth" 312 million—a factor of ten less than "copyright." Even "sex" gets only 876 million hits, in case you're wondering. Don't you think that the concept of "copyright" is a little overdetermined?

The "problem" of copying is not necessarily a legal or ethical one in the strict sense of those words. It cannot be resolved by having people take a stand on either side of a line that says that copying is either good or bad, or that copyright and intellectual-property laws should be supported or abandoned. Such laws have great consequences, and it is necessary that they be debated and addressed—as is being done by legal scholars such as Lawrence Lessig, James Boyle, and Rosemary Coombe. The problem is that there seems to be an almost total lack of context for understanding what it means to copy, what a copy is, what the uses of copying are. A very particular set of philosophical framings of copying, along with the paradoxes, aporias, and internal contradictions that sustain them, emerging out of European and American histories, are now being imposed around the world through globalization and the intellectual-property regimes that accompany it. At the same time, the existence of practices of copying in the premodern West, in the margins of Western culture, and in non-Western cultures has been overlooked, and the fact that different societies have had attitudes to copying that differ radically from our own has been ignored.

I shall argue here that certain non-Western philosophical models of copying, in particular those emerging out of Mahayana Buddhism in its various historical forms, offer us more accurate ways of understanding the diverse phenomena we call "copying," and can help us rethink basic philosophical terms such as "subject," "object," "the same," "different," and "the other"—all of which, depending on the particular ways they've been presented, have historically supported particular cultures of copying. I approach Buddhism not as a historian of religions or a scholar of Asian languages, but as a participant in a set of living intellectual and religious traditions, whose place in the world today is undecided and evolving. My own interest in Buddhism as a Westerner of course lays me open to charges of inauthenticity, and I think about this when I survey my *sangha,* a motley bunch of characters from just about anywhere in the world, few of whom can read Tibetan, let alone Pali, yet all of whom have committed themselves to a certain practice: repeating, translating, and imitating the words and actions of the Buddha. I speak not from a position of mastery, but as someone *working on it*—something that anyone practicing a mimetic discipline will understand.

My goal in this book is to account for our fear of and fascination with copying. I argue that copying is a fundamental part of being human, that we could not be human without copying, and that we can and should celebrate this aspect of ourselves, in full awareness of our situation. Copying is not just something human—it is a part of how the universe functions and manifests. The issue of regulating copying, of setting up laws restricting or encouraging copying, is secondary to that of recognizing the omnipresence and nature of copies and copying in human societies—and beyond. While I do have thoughts on copyright law and intellectual property (IP) as they stand today, the purpose of this book is not primarily to advocate particular changes in copyright law, or to argue that the law as it stands should be disobeyed, or that some kind of free culture that exists beyond the law is possible. I am concerned with what actually is happening, how ideas regarding copying emerge from basic, unresolved issues concerning human consciousness, objectivity, language, and nature. I recognize that law emerges within this framework, and astute readers will quickly note that this book has been copyrighted, and the rights that come with that act are being claimed. I cannot write this book without participating in a wide variety of structures—legal, academic, political, technological, historical, and otherwise—which prescribe the form that my writing takes. But I can easily imagine other ways of organizing and participating in the production and distribution of cultural artifacts such as books, and I'm under no illusions about the limits of copyright law when it comes to how people actually act in this world. Therefore, the book has been issued with a Creative Commons license that allows people to make copies of the book and share it under certain circumstances.

The reciprocal relationship between philosophy and popular, folk, indigenous, sub-, and subaltern cultures proposed in this book may be jarring to some—but one of my goals in writing this book is to generate a description of a popular practice of copying that is not reducible to the legal-political constructions that dominate thinking about copying today. In order to do that, I affirm the existence of a nascent community of people who share in this practice—nascent because the participants perhaps do not yet recognize, literally and otherwise, what they share. In response to contemporary debates about the nature of community, collective, and multitude, I propose that various popular practices of copying already contain a politics that is not only resistant to the dominant logic of late capitalism, but that already operates in a semi-autonomous manner. This practice of copying goes far beyond the banalities of "subculture," or the disturbing impasses of indigenous cultures that find themselves caught between assimilation and forcible insertion into contemporary IP regimes as commodified signifiers of otherness.

In titling this book *In Praise of Copying,* I am seeking to articulate the value of a degraded, devalued part of ourselves and the world around us. As everyone ranging from the artist to the psychoanalyst, ecologist, philosopher, and spiritual teacher is aware, the degraded object, "trash" in whatever form, is a highly potent, energized thing. We create boundaries, separate ourselves from such objects at enormous cost and consequence. Compassion is one of the core Mahayana Buddhist practices. It involves intimacy, tarrying with that which is discursively abjected, in order to learn about it and ourselves, and to see it as it is, free of fear or desire, hatred or grasping. Paradoxically, there has probably never been a moment where an explicit recognition of the power of mimesis has seemed so close: in the hard sciences, the recent discovery of "mirror neurons" places the problem of mimesis at the core of attempts to model cognition; in the social sciences, theories of social contagion, with their attendant popularizations in the form of "memes" and "tipping points," explain the dynamism of human communities in terms of imitation. Are such discoveries merely symptoms of late capitalist ideology, with its own particular appropriations of mimetic power, or do they point to a more fundamental shift that is taking place?

It is not a question of defending everything that is called "copying," every copy that is made. When I started working on this book, I was probably as disdainful as most people of the proliferation of fakeness, and the endless cycles of meaningless repetition that characterize culture. But thanks to the Disney World teaching, I came to recognize that many of the boundaries we have set up between activities we call "copying" and those we call "not copying" are false, and that, objectively, phenomena that involve copying are everywhere around us. Indeed, they are a crucial factor in our ability to make sense of ourselves and the world. I believe that only through recognizing this can we understand the world that we find ourselves in, the world that to some degree we have chosen to make. And many of the most elevated human activities involve copying. In the final analysis, copying itself is neither good nor bad—it all depends on what we use it for, and what we intend with it. But to paraphrase the quote from Gertrude Stein which I have placed at the front of this book: I see it and I love it and I write it.

We live at a time when a radical vision of justice—of the fair distribution of opportunities, possibilities, things, necessities, and luxuries—is lacking. It apparently died with the death of actually existing communist societies, which asserted a vision of another way of doing things, sharing things, even if most of them pitifully failed to live up to this vision. Copying seems to manifest as a pressing issue at moments where there is a radical shift in societies. The Statute of Anne of 1709, the first copyright law, was in part a rearguard effort to protect the rights of the Stationers Company in the face of the effects of the English Revolution; copyright and patent law was inscribed in Article 1, section 8 of the U.S. Constitution (1787), and in a law of 1793 in France; the Russian Revolution was accompanied by a variety of changes to copyright law (which had hitherto been in line with those of bourgeois European law), including a 1923 decree nationalizing the works of authors such as Tolstoy, Gogol, and Chekhov.[1] Why should this be so? Not necessarily because at such moments more people are engaged in acts of copying or the production of copies, but because the ideologies that sustain the illusion of the permanence and naturalness of a particular society disintegrate, revealing the various processes which actually sustain such societies. Such as imitation. Clearly, we are living through such a shift right now, but without any particular sustaining vision of what lies beyond. This is not necessarily a bad thing. Rather than coming up with a new illusion and trademarking it, we have an opportunity to recognize what our situation consists of. "Copying" is just one word in one language, an apparently trivial matter—yet, for reasons which I will explore, an activity that exceeds itself in every way, opening up to a vastness as surprising as it is undeniable.

1/What Is a Copy?

> IT'S A VUITTON
> and you know that Vuitton trunks have been called "the trunks that last a life-time."
> A VUITTON WARDROBE TRUNK
> not only IS French but it LOOKS French, not only IS the finest but APPEARS to be the finest.
> VUITTON TRUNKS ARE GENUINE!
>
> —Advertisement in *Town and Country*, May 15, 1922, quoted in Paul-Gérard Pasols, *Louis Vuitton: La Naissance du luxe moderne* (2005)

Louis Vuitton

Brooklyn, New York, April 2008. A row of street stalls in front of graffiti-covered iron gates. Tables full of merchandise: Louis Vuitton handbags and wallets, with their familiar "LV" monograms; brown and beige; white with multicolor fruit-like designs. You can find them for sale on Canal Street in New York, in the night markets of Hong Kong and Singapore or the covered market in Mexico City, and in many other places around the world where the urban poor go to shop—"LV" articles piled up alongside the Patek Philippe watches, Chanel perfume, North Face jackets, and Adidas shoes. Copies, fakes, counterfeits; cheap, poorly made reproductions . . . or are they? For you are not in a night market, or on the street. You are standing inside the Brooklyn Museum, surrounded by cameras and elegantly dressed men and women; Kanye West is performing in another room in the building. This is the opening night for Copyright Murakami, a retrospective devoted to the work of Japanese visual artist Takashi Murakami, including his celebrated collaborations with Louis Vuitton, such as the multicolor monogram handbag you just saw. And the bags in the street stalls are the real thing, made by Louis Vuitton, for sale at high prices. According to spokesmen for the company, the fake street stalls selling fake fakes are intended to draw attention to the phenomenon of counterfeiting, the production of illegal copies of Louis Vuitton's products.[1]

Vuitton handbags have been called the most copied objects in the world.[2] This statement, part of the folklore of contemporary global consumer culture, seems immediately open to question. Louis Vuitton, after all, is a manufacturer of luxury goods which are defined, even in this age of global branding, by their scarcity. Internet folklore has it that only 1 percent of Louis Vuitton bags are actually made by the company.[3] The copies, then, would be the 99 percent made by others. The selling of such mass-produced copies—which in its current form can be dated back to the 1970s, when Vuitton bags began to be made en masse in various East Asian locations—is not a new thing. In fact, Vuitton's famous "LV" monogram was developed in 1896 by Louis Vuitton's son Georges, as a trademark that would authenticate the family firm's products, in response to the alleged copying of Vuitton Senior's checkered-cloth design. Although Georges designed the monogram to distinguish his company's products, today it is the distinctive "LV" logo that makes the bags so easy to copy.

The market for such copies has developed in surprising ways. Today in Taiwan, we are told that there are five grades of copy, ranging from the highest—which are handmade, almost indistinguishable from the bags made by Vuitton, and costing thousands of dollars—to the cheap plastic fakes available in night markets in cities. Some of these bags, which are sold complete with certificates of authenticity, fake receipts, and logo-stamped wrappings, have been "returned" to stores which sell the real items but which did not detect the replicas. On the other hand, famous movie stars have been spotted carrying Vuitton bags which include designs that are not actually made by the company.[4] Furthermore, because of the difficulty in actually purchasing some of the limited-edition bags made by Vuitton and other companies such as Hermès, with its famous "Birkin" bag, it has become fashionable to celebrate rather than hide the fact that a bag is a copy, and the vogue for certain copies has resulted in their prices exceeding those of the originals that they supposedly imitate. Online, one can find images of Vuitton bags which bear the word "FAKE" in bold letters on the side of the bags.[5]

The fragility of the trademark as an identifier of authenticity is illustrated by the fact that in China destruction of copies is often prohibitively expensive, and so labels from counterfeits are merely removed and the now-generic items sold in the marketplace again.[6] Conversely, in order to circumvent the law on illegal vending of counterfeits in Counterfeit Alley in New York, fakes are often sold as "blanks" in one location, with logos and other trademarks being added at a second location later.[7] The instability of the word "copy" in this situation is also illustrated by the fact that factories that produce "originals" under outsourcing contracts from international businesses may also produce the same goods illegally on the "ghost shift" at night, which are then sold as fakes or counterfeits.[8]

The ironies on the Vuitton side mount, too. The "LV" monogram was designed four years after Louis Vuitton's death. The firm remained a family business for many years, but became a publicly traded company in 1984; the family lost control of the business in

1990, after a hostile takeover bid by Bernard Arnault that resulted in the formation of the "French" luxury conglomerate Louis Vuitton Moët Hennessy (LVMH). This shift was magnified by the hiring in 1997 of New York–based fashion designer Marc Jacobs as the brand's artistic director and the hiring of global talent such as Murakami to develop product designs for the company. Although the company still makes luxury hand-crafted goods, it currently has 390 stores around the world. Unlike many other luxury businesses, Vuitton has resisted the urge to outsource production of its goods, maintaining fifteen factories in France; but the company also recently opened factories in Spain and the United States, and began a joint factory venture in Pondicherry in India. So Vuitton is a mass-producer of luxury, artisanal, unique individual bags, faking the faking of its own products at an art exhibition, while zealously pursuing the prosecution of the actual fakers through police action and courts of law around the world.

The not-by-chance meeting of Murakami and Vuitton in an art museum in Brooklyn embodies many of the contradictions involved in thinking about copies. Murakami is one of the most famous visual artists working today, exhibiting his paintings, the pinnacle of individualistic self-expression, in art museums, the most prestigious archives of the unique and original object. In the 2008 Brooklyn show, there was a Louis Vuitton boutique where the visitor could purchase some of the handbags Murakami designed in collaboration with Vuitton. A number of the paintings in the exhibition featured Vuitton's logo incorporated into their complex "superflat" surfaces.[9] At the entrance to the Copyright Murakami show, visitors were greeted by the statement: "The concept of copyright holds an exalted position within Murakami's practice, rooted in the acknowledgment of his work as simultaneously interweaving deeply personal expression, high art, mass culture and commerce." The title of the show references a long-standing stereotype concerning the illegal and anonymous production of copies in East Asia, and playfully transforms it. Murakami himself runs a company called Kaikai Kiki, which manages artists and produces and sells merchandise. At the same time, his own work is based on an explicit appropriation of materials from a variety of sources, including traditional and contemporary Japanese culture. Furthermore, the idea for the museum installation itself appears to have been copied from previous works, such as an installation by Fred Wilson at the 2003 Venice Biennale in which he hired a black man to stand outside the main pavilion selling fake generic designer bags, and Korean artist Zinwoo Park's 2007 exhibition of real Louis Vuitton "Speedy" bags with the label "FAKE" attached to them.[10]

The everyday saga of intellectual property and its protection is here elaborated to an unusual degree. Marc Jacobs may claim that the Brooklyn Museum's tableau was just a little amusement, but the fact that all the players involved choose to pay close attention to such an apparently trivial matter as copying should indicate the existence of a crisis. Such a crisis might involve: the globalization of commerce and the transport of texts, images, symbols, objects, and products across national boundaries and cultural spaces in a way that calls into question the ownership of such things; the problem of when some "thing" can be called "art" and the ever-expanding role of the museum in legitimating objects as being art or otherwise, even as museums themselves are forced to function as part of a market economy; consequently, the erosion of the gap between financial and aesthetic value and the increasingly open question as to the source of the prestige of particular fabricated objects; the inability of the law to resolve, both intellectually and practically, questions about the identities of objects, about what can be claimed as private property or not, and what the rights of various parties as to the use of things are; last but not least, the apparent indifference of the general public to whether the things that they buy are "real" or "fake," "original" or a "copy," as evidenced by the expanding market for both originals and copies of many products.[11]

So: what exactly constitutes a "copy" in this situation—or rather, what does not? Writing admiringly of the LV copies available in New York City, for example, fashion journalist Lynn Yaeger struggled to put her finger on the difference between an original LV bag and a well-made copy.[12] The site Basicreplica.com, one of a number of Web-based companies that in 2009 offered high-end copies of Vuitton, along with Dior, Marc Jacobs, and others, proclaimed:

> No tongue in cheek, we can honestly say that our Louis Vuitton replica bags are absolutely indistinguishable from the originals. You can take your Louis Vuitton replica handbag to a Louis Vuitton flagship store and compare, feel the leather, test the handles, check out the lining—not even a Louis Vuitton master craftsman will be able to tell which is the original and which the Louis Vuitton replica handbag from Basicreplica.com. Louis Vuitton replica bags with the same Alcantara lining, quality cowhide leather given a finish that oxidizes to a dark honey just the way the original Louis Vuitton handbags colour as they age, authentically original imitations of the real originals![13]

Aside from being a fabulous rhetorical flourish, what is an "authentically original imitation"? Or more specifically: What is a copy? In everyday parlance, the word "copy" designates an imitation of an original—for example, a copy of a Louis Vuitton bag. But a brief survey of the kinds of objects called "copies" today raises basic questions about this definition. What does it mean to say that something is a copy of something else? How is the claim that object A is a copy of object B established? What do we mean when we say that A is "like" B, that it imitates it? At first, these questions strike one as banal and the answers obvious or self-evident. But when original and copy begin to overlap to the extent that they do today (and the struggle to maintain the distinction between these two things, "original" and "copy," is precisely what constitutes the crisis, to my mind); when original and copy are produced together in the same factory, at different moments; when a copy is actually self-consciously preferred to the original, we must ask again: What do we mean when we say "copy"?

The Platonic World of Intellectual Property

What is the origin of the vocabulary—legal, commercial, aesthetic, or otherwise—that is used to describe the complex global situation of the Louis Vuitton bag? To answer the question adequately might require one to tell a history of the world, which is perhaps why no one has attempted it. Nevertheless, it is a situation in which a specific philosophical history is being deployed, knowingly or not, ingenuously or not, by all those involved. In this history, Plato's writings on *mimesis*—a word usually translated as "imitation" but also "copy," "representation," "reproduction," "similarity," or "resemblance"—play a key role.[14] In Plato's *Republic,* Socrates presents the argument that everything in this world is an imitation, because it is an echo or reproduction of an idea that exists beyond the realm of sensible forms. A Louis Vuitton bag is the imitation of an idea, in leather and other materials, while a photograph of such a bag is an imitation of an imitation. In what way is the bag an imitation of an idea, though? In an analysis of the Platonic idea, Martin Heidegger gives an answer to this perplexing question: "Mimesis means copying, that is, presenting and producing something in a manner which is typical of something else. Copying is done in the realm of production, taking it in a very broad sense. Thus the first thing that occurs is that a manifold of produced items somehow comes into view, not as the dizzying confusion of an arbitrary multiplicity, but as the many-sided individual item which we name with one name."[15]

So copying is a matter of "presenting and producing something in a manner which is typical of something else." All copies are made—they are produced—and the making involves an attempt to turn something into something else, so that that which is produced is now "like" something else. But in what way is it "like" something else? Why is the bag "like" the idea of a bag? Or for that matter, why is the fake Louis Vuitton bag sold on Basicreplica.com like the original object sold in Vuitton's Paris flagship store? Heidegger responds: "Making and manufacturing . . . mean to bring the outward appearance to show itself in something else . . . to 'pro-duce' the outward appearance, not in the sense of manufacturing it but of letting it radiantly appear" (176).

Outward appearance is crucial here, for "in the outward appearance, whatever it is that something which encounters us 'is,' shows itself" (173). It is outward appearance that makes something "like" something else; but more profoundly, it is in outward appearance that the idea, the essence of something, shows itself. The quote from a 1922 Louis Vuitton ad that figures at the head of this chapter articulates this Platonic belief very clearly: the bag not only "looks like" something but "IS"; it not only "IS" but "appears." The famous "LV" logo also makes sure we know that something not only "appears" to be an actual Louis Vuitton bag, but "IS."

The astute reader or shopper will immediately realize that there is a problem: the fact that something appears to be a Louis Vuitton bag does not mean that it is. For, as we know, an "LV" logo, indeed the entire design of a Louis Vuitton bag, can be copied. Plato, too, recognized this problem, and Socrates poses the following riddle to his respondents in order to think it through: There exists a producer who can produce not only chairs or tables, but the sun, mountains, everything in this world. Who is this producer? Answer: Someone holding a mirror. In the mirror, everything in the world is produced and appears. Again, we ask, in what sense does a mirror "produce"? Heidegger explains that if we understand "produce" to mean manufacture, then obviously a mirror cannot be used to manufacture the sun. But if we understand "produce" to mean "manifest the outward appearance of," then a mirror does "produce" the sun, even if it clearly does not manufacture a sun.

There are, then, different ways in which an outward appearance can be produced—and different producers, too: the god produces the idea, the craftsman is able to make the idea radiantly appear in an object, and the painter makes it appear in a painting.

What then differentiates these three ways of producing outward appearance? The latter two are diminutions or distortions of the first. Hence Plato's mistrust of mimesis, and of the artist—the mirrored image, and even the craftsman's object, confuse the ignorant as to what is essential. At the same time, it is the Platonic belief that the outward appearance of something indicates its essence which continues to generate much of our confusion about what a copy is. When we say "an original," we usually mean something in which the idea and the outward appearance correspond to each other. There is no distortion in the relation of appearance to essence, to "what a thing is." Copies, then, for Plato and for us, most of the time are distortions of this relationship. The mirror produces the sun, yet it is not the sun. Basicreplica.com produces a Louis Vuitton bag, yet the article is not a real Louis Vuitton bag.

Under "Frequently Asked Questions" on the website, the people at Basicreplica.com deftly exploited the confusions that underlie Platonic thought:

"*1. Are these Authentic Louis Vuitton hand bags?* No, we do not sell Louis Vuitton registered trademark bags. The real Louis Vuitton bags can only be bought from authorized dealers. Our bags are replicas. They have all of the proper labeling in all the correct places, lining, locks, and keys, are of the high quality you should expect, and look authentic."[16]

The bags are not authentic; they are replicas. But they look authentic. What is the difference between something "looking authentic" and "being authentic"? Especially if, taking Basicreplica at their word, we can say that everything in the copy is made with the same materials and is of the same "high quality." If the 1927 Louis Vuitton ad claimed that LV bags not only have essence, but look as though they do—their outer appearance being in accord with their essence—then Basicreplica could claim that although their bags' outward appearance was identical to those made by Louis Vuitton, they were not liable to charges of copyright or trademark infringement, because they were not claiming that the bags were "Authentic Louis Vuitton hand bags."

Intellectual-property law functions through Platonic concepts. IP law's three constituent parts—copyright, trademark, and patent law—are each built around the paradox that you cannot protect an idea itself, but can protect only a fixed, material expression of an idea. One claims an idea as property by materially fixing it through describing a process for realizing it (patent law), by inscribing or figuring it materially in the form of a picture, text, notated music, film sequence (copyright law), or by developing some method of inscription that one uses to mark otherwise generic objects as one's own (trademark law). What is the ontology of intellectual property?[17] Ideas cannot be owned, because they are intangible, but the original expression of an idea can be owned when it is tangible, material, fixed. While the idea itself exists in a realm beyond the human realm, the expression belongs to this world, and to the person who, receiving the idea as author, inventor, or owner, fixes it materially as self-expression through his or her labor and turns it into property. This is called "originality." Others who fix it materially via access to the this-worldly original expression, rather than receiving the idea, are said to be making a copy. The law protects the rights of the former, but not the latter—unless the expression is a fact, a generic term, etc., in which case it belongs in the public domain.

In the age of globalized capital, the commodity itself has adapted to the structures of Platonic legal ontology. Manufacturers work to produce products with distinctive outward appearances that fix, mark, the originality with which they claim to express an idea. Thus, the distinct shape of Louis Vuitton's Monogram bag can be copyrighted, the name "Monogram" and the inscription "LV" on a bag can be trademarked, and certain innovations in the otherwise generic product called a "bag" can be patented. And those who wish to make similar products must situate their productions within certain legal spaces: that of the art object, protected by fair-use doctrine (though Vuitton has attempted to prevent artists from making LV bags for this purpose without the company's permission, at the same time legitimizing the productions of others such as Murakami or Stephen Sprouse, with whom the company is collaborating); the parody (for example the "Chewy Vuiton" squeaky toys made by pet toy manufacturer Haute Diggity Dog, which Vuitton unsuccessfully attempted to sue);[18] the generic item called a "bag" which receives no IP protection; or the more spurious, yet also more philosophical arguments offered by Basicreplica.com. At all costs, one should avoid being associated with copies or copying, or face being banned from the republic! It all comes down to what "is," or rather what is legally granted the status of being. Yet paradoxically, since ideas do not or cannot receive legal protection, IP law encourages those who produce commodities to exaggerate the inevitable distortion of the idea as manifest in the actual object. And the result of this is the kitsch version of originality, "thinking outside the box," that prevails in the marketplace today.

Alternatives to Platonic Mimesis

All of this assumes that the Platonic model is true. It is unclear how seriously the producers of the Basicreplica.com website—or the advertising agency that produced the 1927 Louis Vuitton ad—take their astute deployments of Platonic concepts. Platonism, as new-media theorist McKenzie Wark recently pointed out, is a game, complete with screens, darkened rooms, and headsets. Through the immense historical networks which have resulted in globalization, the game has been installed (to use the word explored by theorist Philippe Lacoue-Labarthe in describing the advent of particular mimetic regimes) almost everywhere today, and in a limited sense this game is functional. But beyond this limited sense, with its official protocols of exchange, law, ownership, and identity, what accounts for the multiplication of Louis Vuitton bags?

The history of the Western philosophical tradition, beginning with Aristotle, consists in good part of a series of responses to—modifications, negations, and reversals of—Platonic mimesis.[19] An in-depth review of this tradition is beyond the scope of this book and I refer the reader to the excellent accounts that are available.[20] Christianity takes up Platonic ideas in a variety of ways, from Augustine's positing of the world as a "region of dissimilarity" separated from God, to Aquinas' *Imitation of Christ,* in which mimesis has a positive valence as a way of participating in the divine.[21] Although, after the Renaissance, mimesis thus named is increasingly downplayed in Western philosophy, the underlying problematic of mimesis remains.

As for contemporary critical theory, we can summarize the situation as follows. Elaborating on Nietzsche's "reverse Platonism," Gilles Deleuze observed that the Platonic Idea is always accompanied by a swarm of simulacra, fakes, and copies that threaten it, distort it, etc.; and he affirms the equal ontological rights of these simulacra. Jacques Derrida, continuing Heidegger's critique of Western metaphysics, tracked down residual traces of Platonic idealism in Husserl and others, proposing the freeplay of the trace as an alternative way of understanding phenomena. Michel Foucault, in "What Is an Author?" argued that authorship and the language of original and copy that accompanies it are variously constructed by particular legal-social-political regimes. Thus, we find ourselves in the now-familiar condition of postmodernity, set out most famously by Baudrillard in *Simulations:* a world of "copies without originals."

From the perspective of this tradition, we see an infinite proliferation of Louis Vuitton bags, regulated by something like Foucault's "author function," and the historical-social-political institution of a system of property rights management that assigns ownership and authorship to an entity known as Louis Vuitton. But while this is a valid description of the situation as far as it goes, it does not explain how something like a Louis Vuitton bag comes to appear as such at all. While it affirms the power of the fake to challenge the original, through a reversal of Platonism, it does not explain how this happens—or why it fails, insofar as it does. While "Platonic mimesis" is disavowed, it insidiously reasserts itself in the absence of persuasive alternatives.[22]

We find ourselves in a certain impasse—legally, philosophically, theoretically. But the various philosophical and theoretical responses to Plato hardly exhaust the possible framings of mimetic phenomena. And given the situation of the vast swirl of objects known as "Louis Vuitton bags" circulating around the planet today, it would be impossible to claim that this situation is solely the result of a particular history, or a competing set of counterhistories, within Western philosophy. This is confirmed by a number of recent comparative studies revealing that historically, philosophically, and otherwise, different cultures have had very different ways of understanding and valuing the phenomena that today are associated with intellectual property.[23]

While it is no doubt the case that there are traces of other traditions, other practices, in European and American cultures and elsewhere, that appear, contest, deform, and transform these bags, I would like to raise a more fundamental issue here. For if we can agree that there are no Platonic essences, the challenge that remains today in terms of a philosophical meditation on copying is to understand how essencelessness actually functions—how something like a world in which originals and copies appear actually takes shape. In this regard, a number of Asian philosophical traditions have elaborated complex and relevant ways of thinking essencelessness in regard to phenomena. Indeed, it can be shown that the modern philosophical critique of essence, from German Idealism to existentialism, has occurred in dialogue with, and in response to, Asian philosophical traditions.[24] Much of Heidegger's later interest in the overcoming of metaphysics bore the (mostly unacknowledged) influence of his studies in East Asian philosophical traditions, including a decades-long conversation with members of the major modern Japanese philosophical school, the Kyoto School. Adherents of the Kyoto School, along with a number of modern philosophers, have made detailed critiques of Platonism and Western metaphysics from a variety of traditions and philosophical perspectives; yet these critiques, along with those of Western Buddhist philosophers, are rarely responded to within contemporary critical theory or philosophy.[25] Furthermore, there is evidence of the passage and transmission of philosophical thought between Europe and Asia as far back as 500 B.C., which would be both the period of the pre-Socratics and of the Buddha—meaning that Asian influences on Plato's philosophy, and vice versa, cannot be ruled out.[26] Thus, the philosophical question I am asking concerning Louis Vuitton is at once interior and exterior to that of the Western philosophical tradition as it is usually defined.

A kind of philosophical apartheid prevails in which a "modern" critical approach must remain separate from any philosophical tradition whose source lies outside the narrowly defined canon of Western tradition—and it is this separation that helps to maintain a discourse of copying that otherwise would surely fall apart. The legacy of Orientalist appropriation and the stereotyping of Asian and other societies remain a matter of concern, but it is a mistake to disavow the possibility that there are other philosophies and practices of copying, in order to avoid associating particular individuals, cultures, and societies with "criminal activities" that are "antimodern." Recent scholarship on alternative modernities—which would include the concept of "Buddhist modernisms," in the phrase of Donald Lopez—argues for the dynamism of such traditions as they encounter modernity. Furthermore, as we know from the work of Bruno Latour, the concept of the modern itself is open to question, and one of the consequences of this questioning is the validation of a variety of ways of knowing and constellating a world. It may seem extravagant to ask for a renegotiation of what is called "philosophy" all for the sake of a Louis Vuitton handbag—and possibly a fake one at that—but that is what I am proposing here.

Śūnyatā and Copying

The topic of essencelessness in Buddhism is a vast one, and nuances in the way that the topic is responded to have been determinative in the founding of particular schools and traditions. *Anattā,* or "no-self," is already a key concept in the Pali sutras, which are generally considered the earliest extant written records of the discourses of the historical Buddha—as is "dependent origination" (Pali: *paticcasamuppāda;* Sanskrit: *pratītyasamutpāda*). However, the critique of essence (Sanskrit: *svabhāva*) comes to the fore in foundational Sanskrit Mahayana texts such as the *Prajñāpāramitā Sutras,* the *Laṅkāvatāra Sutra,* and the writings of Nāgārjuna, founder of the Madhyamaka School, and remains a key component of many existing Asian Buddhist traditions, notably in the Tibetan diaspora and East Asia.

Mahayana Buddhism is concerned with the notion of essencelessness or emptiness (Sanskrit: *śūnyatā*), as regards both subjects and objects.[27] Contrary to the stereotypical nineteenth-century European view, the critique of essences in Buddhism does not result in a nihilistic or languorous dismissal of the phenomenal world (although "nihilism" is also an accusation that Buddhists from various traditions have made concerning traditions other than their own, Buddhist and non-Buddhist).[28] Rather, Mahayana Buddhism seeks to account for the way in which the phenomenal world *appears* to us, and to establish the true nature of this appearance. In the view of the Madhyamaka school of Nāgārjuna, "copying" in its Platonic form would emerge out of the belief that there is an original object with an essence that could be copied; and this belief could be logically refuted. For if objects really did have essences, there could be no copying of them, since that which one would make the copy out of would continue to have its own essence, and could have only this essence, rather than that essence which is implied by the transformed outward appearance that would make it a copy. Similarly, if the essence of a thing were truly fixed, it could not be transported to the copy, and imitation, even as a degradation of the original, would not be possible.

The Madhyamaka philosopher's response to the Platonic doctrine of the idea would be to ask the Platonist where one can find the ideal form which supposedly constitutes the real Louis Vuitton bag, and, through the systematic negation of all the possibilities, to demonstrate that it has no existence. We can find nothing but the bags that are around us, some of which we call and designate "Louis Vuitton bags." This designation is always necessarily a relative one. The Louis Vuitton bag does not know it is a Louis Vuitton bag, even if it has "LV" inscribed on it. To a person from the tenth century, to a dog, or to a bacterium, the designation "Louis Vuitton bag" would be meaningless, as far we know. There is no essence to the bag which guarantees that it is recognizable as such. This is not to say that the designation "Louis Vuitton bag" is always meaningless—but the meaning is contingent, relative, dependent on causes and conditions: I must be able to speak some English or French; I must have eyes or other senses that are able to sensually register the bag; preferably I will live in a society in which I have already been educated as to the meaning of these words, this object. Thus, if the Madhyamaka philosopher asserts that the Louis Vuitton bag is empty (i.e., *śūnya*), this does not mean that it is nothing, or that there is nothing inside it; it means that the statement "this is a Louis Vuitton bag" is a relative and contingent one, dependent on an act of designation or labeling. And from an absolute point of view, there is no possibility at all of fixing, defining, or characterizing the object that is designated "bag."

There has been considerable debate historically among different Buddhist schools as to how the mechanism of designation works, leading several contemporary scholars to propose a complex set of Buddhist ecologies of the sign.[29] But in each case, designation is an unstable, impermanent act by which something temporarily appears to appear. So it is with the famous Louis Vuitton monogram logo, which, although it says "Louis Vuitton," was actually developed by Vuitton's son Georges four years after his father's death, thus marking an absence rather than the presence of "Louis Vuitton." When Georges brought the "counterfeiter" of his father's design to court, the counterfeiter brought forth an old cloth manufacturer's book to prove that his father had in fact copied "his" design from someone else;[30] thus, the original that Georges sought to defend turns out to be a copy, too. Furthermore, it has been claimed that the monogram logo was itself copied from a variety of sources, including Japanese heraldic designs (a no doubt desirable indexing of the logo in terms of growing the Japanese market) and the wallpaper in the Vuitton family kitchen. And so it is with all the other marks of authenticity of an LV bag. The packaging the bag comes in, the tags attached to the bag, even the receipt indicating that the bag was purchased at an actual Louis Vuitton store—all are subject to the same limits. They are acts of designation, rather than guarantors of essence; as such, they are impermanent and they can themselves be copied. It is the emptiness of all phenomena, their lack of essence, which makes copying possible; but more important, this emptiness is what makes it possible for anything to appear at all.

Sameness and Nonduality

The deconstruction of all claims of conceptual identity, and the revealing of the impermanence, provisionality, and relativity of all names and forms, does not in itself take us any further than the postmodern position, in terms of understanding phenomena of imitation. Copying requires the recognition of a similarity between two things; but without essences, how could there be such a similarity? How could one say that two things are the same? We can speak of "designation" or "construction"; but how do such acts function, insofar as they do—and why?

Sameness is an uncomfortable but decisive topic in contemporary theory and philosophy—precisely because of the Platonic legacy, and the history of the imposition of white male European-American hegemony through a set of supposed universals, along with various attempts at overthrowing or reversing this hegemony which often merely repeat it. Sameness appears in Derrida's famous essay "Differance," where he argues, citing Deleuze, that "philosophy lives *in* and *from* differance, that it thereby blinds itself to the *same*, which is not the identical. The same is precisely differance (with an *a*), as the diverted and equivocal passage from one difference to another, from one term of the opposition to the other." Derrida goes on to describe the construction of philosophy as "the systematic ordering of the same."[31] But this sameness was not pursued in poststructuralist thought, and "differance" slipped back into a mere, reified "difference" purged of the nondifference with which, according to the most econstructive practice, it must be coextensive. More recently, ı has written that for ethics "the real question—and it is an ex- inarily difficult one—is much more that of *recognizing the Same.*"[32]

The most elegant and concise formulation of sameness in the Western tradition occurs in the work of Walter Benjamin, in his discussion of similarity: sensuous and nonsensuous. In his essays "Doctrine of the Similar" and "On the Mimetic Faculty," he describes "the commonplace sensuous realm of similarity" as the kind that we are familiar with, Platonic or not. Benjamin also calls it the "semiotic element." It is concerned with semblance, "outward appearance," "likeness and so on." "Nonsensuous similarity" (German: *unsinnliche Ähnlichkeit*) is described by Benjamin as that which brings together a word and the object it designates, and makes them connected, "similar." It is, says Benjamin, what connects spoken word to written word, and makes handwriting analysis possible. The term "nonsensuous" is used in medieval Christian theology (for example, in the works of Meister Eckhart and Nicholas of Cusa) to indicate the intelligible (i.e., the realm of ideas) as opposed to the sensuous, but Benjamin gives the term a materialist twist. He comes up with several examples of nonsensuous similarity, but the term remains enigmatic, and I propose to reframe it according to the Buddhist schema that I have just set out.[33]

In a preliminary study for "On the Mimetic Faculty," Benjamin writes: "The communication of matter in its magical community takes place through similarity."[34] What is this magical community of matter? Is it magical merely in the sense of being irreducible to words? No—if we reverse the order of the quotation, similarity happens through the communication of matter, through its community, which is interrelated. It is interrelated in being primordially undifferentiated. Differentiation, meaning the cognitive processing of the sensual and the intelligible, establishes separation, which establishes separate forms. But in the terms of Mahayana Buddhist philosophy, differentiation is itself possible only within the context of the interdependence of those semiotically "similar" things that provisionally appear to us as separate entities. In Benjamin's terms, sensuousness separates the magical community of matter into "things" which are semiotically similar or dissimilar, but only the coemergent nonsensuous similarity of all provides the necessary conditions through which sensuous similarity can be recognized.

The term "nonsensuous similarity" is, as we have noted, awkward. The kind of similarity that Benjamin is talking about would have to be called "nonconceptual" as well as "nonsensuous," so as not to be misinterpreted as a metaphysical or transcendental substrate of the sensuous. In his writings on hashish, Benjamin says one should scoop sameness out of reality with a spoon, as the source of great happiness.[35] What Benjamin means by "sameness" is precisely nonsensuous, nonconceptual, nonsemiotic similarity. In Buddhism and Vedanta this sameness is called "suchness" (Sanskrit: *tathātā*), and this suchness is beyond notions of semiotic sameness and difference. It is this particular sameness that in Benjamin's terms, "flashes up" through the "semiotic element" or, in Buddhist terminology, appears in/as relative, interdependent cognitive and phenomenological structures.

Thus, "sameness" in this context does not mean a substrate that underlies all the permutations of difference. It is not merely the reified nothingness with which Hegel dismisses Buddhist approaches to the Absolute, or another way of saying "monism."[36] Nevertheless, there is also an intense and ongoing debate within various Buddhist traditions as to whether the emptiness of all phenomena can be described as a sameness, and if so, how one can speak of it without falling into reifications of various kinds which would reintroduce an essence as a hidden substrate. In Mahayana Buddhism, the distinction between the Yogācāra and Madhyamaka philosophical schools arises in part in relation to the former's belief in a groundless mental ground called the *ālayavijñāna,* an undifferentiated ocean that all differential designations emerge from and fall back into. The Madhyamaka assert that any such figuration necessarily reifies emptiness into a thing with an essence, and that absolute emptiness is completely beyond all conceptual formulations. These subtle philosophical distinctions regarding the relationship of sameness to emptiness have been highly significant in the history of Tibetan Buddhism, and the debate between different schools of thought remains unresolved.[37]

A number of recent comparative studies of Buddhist philosophy and contemporary critical theory have focused precisely on this problem, all arguing that Buddhist philosophy addresses and potentially corrects certain flaws in modern theory, concerning the relationship between essencelessness and phenomena.[38] In a comparison of Chinese and Indian Mahayana texts and Derridean deconstruction, Youxuan Wang sees in all of them a destabilization of the hierarchy of the sign that reveals an infinite, interdependent, groundless chain of significations (the union of dependent origination and emptiness in the former; the freeplay of traces in the latter). Wang calls this groundless ground a "reversed mimesis" in which, paradoxically, every appearance of a sign is a mark of emptiness. Graham Priest and Jay Garfield, in their work on the history of philosophical paradox and contradiction, identify this as "Nāgārjuna's Paradox": "If things lack fundamental natures, it turns out that they all have the same nature, that is, emptiness, and hence both have and lack that very nature."[39]

All acts or events of signification are "the same," then, since they point to their own lack of essence. At the relative level, signification functions through the chain of signifiers to reveal the relative world of appearances. But every signification also paradoxically "signifies" emptiness. Thus, difference and sameness are neither different nor the same; and what is—i.e., what has the ontological status of truly existing—is emptiness itself.[40] Emptiness, then, has a double status, of relative and absolute truth. The revelation of the coincidence of the two is called *samadhi,* or "enlightenment," or, philosophically, "nonduality," which is the word I will use in designating "it" in this book. Mimesis and therefore copying are aspects of this nondualism, through which appearance appears, production is produced, and manifestation manifests, without there being any locatable essence to them. I unapologetically insist on the value of what Robert Magliola pejoratively terms a "rhetoric of holism," in contrast to his preferred "Derridean differentialism."[41] For it is precisely a discursive meditation on nondual sameness or suchness that allows us to understand the appearance of mimetic phenomena.

Things That Have Touched

Copying, then, consists of a series of practices that "magically" work with a recognition of fundamental nonduality, in order to manipulate appearance. In his recent work on mimesis, anthropologist Michael Taussig reassesses the work of earlier anthropologists on magic and mimesis, particularly that of Sir James George Frazer, who, in his vast tome *The Golden Bough,* pared magic down to two effective Laws: that of similarity and that of contact, or contagion. The law of similarity means semblance or likeness, and we have explored it above. Taussig discusses the example of voodoo dolls and other ritually potent mimetic figurines, and Freud's observation in *Totem and Taboo* that voodoo dolls often bear little resemblance to the people they are supposed to be effigies of. Taussig concludes that the effigy "is not a copy—not a copy, that is, in the sense of being what we might generally mean when we say a 'faithful' copy."[42] Taussig argues that it is the addition of body parts such as hairs, skin, or bodily secretions to the effigy which is decisive in the construction of an effective effigy. And this is the law of contact—that "things which have once been in contact with each other continue to act on each other at a distance after the physical contact has been severed" (52–53). Taussig goes on to say that most examples of sympathetic magic, based on likeness, turn out to have a component of contagious magic to them, too; so that, as with the example of a fingerprint, it finally becomes impossible to disentangle the aspect of substance from image, contact from likeness. Advertising, in particular, is driven by this magic. A famous 2005 ad for Louis Vuitton features the actress Uma Thurman lounging on a stone structure, her arms above her head, her body somewhat exposed but her hands coiled around a monogram bag which sits in front of her. Thurman looks at the camera with a casual but powerful neutrality, the whole image radiating a strange mixture of exhibitionism, confidence, and security. By touching the bag (the gesture is repeated again and again in Vuitton campaigns), Thurman confers on it the power of her celebrity. Her likeness is transmitted by photograph and print. The consumer purchases the bag, and the magic charge of celebrity is transmitted to him or her, too.

Chinese art history texts speak of a Tao (or Way) of copying, and copying was considered one of the six elements of painting. The ability to copy texts or images on scrolls was prized, but the manifestation of "spirit vitality" was prized in copying, beyond verisimilitude. T'ang writer Chang Yen Yuan notes that "the representation of things necessarily consists in formal likeness, but likeness of form requires completion by a noble vitality. Noble vitality and formal likeness both originate in the definition of a conception and derive from the use of the brush."[43] This spirit vitality is an aspect of nonduality, and beyond likeness; it is marked by the touch of the brush on paper, style made manifest in a moment of tactile engagement of hand, ink, brush, and paper.

The law of contact, thus described, provides us with another way of thinking about nonduality. For what exactly is it that is contagious in matter which has passed from a person to an effigy and back again? After all, only tiny traces of the original are introduced into the copy. It is the assertion of the nonduality of *a* and *b,* through the touching of these apparently separate entities. What would that be? The notion of "bonding," or "binding," can help us here. The word is a key one in the Western hermetic tradition—for example, Renaissance philosopher/heretic Giordano Bruno puts it at the center of his theory of magic.[44] It also appears in a striking variety of ways in Asian religious traditions: *yoga* is derived from *yug* or "yoke," *tantra* is a "weave," *bindu* is a "point," but notably the point where the nonsensuous and the sensuous converge. Bonding indicates the particular direction in which mimetic energies are configured and directed. This is important because it suggests a possible response to an obvious objection to Taussig: that not everything that touches something else becomes it, and, for that matter, not everything that looks like something else is considered a copy. Given the impermanence of forms, what is it that holds them, temporarily, provisionally, in particular forms, identities, and configurations? Bonding indicates a set of intentions, practices, and structures that work to produce the experience of subjective and objective things, including copies. Through bonding, forms and entities are temporarily and provisionally manifested as limited, and as separate and "different" within the field of nondual awareness. Bonding is not merely "the semiotic," but the particular forms that semiotic constellations take, and it is thus an eminently political and historical process. The scandal of copying is the revelation of this process of bonding—and unbonding, its necessary counterpart—as pervasive and insistent, against all claims of permanence and essence.

René Girard has a particular take on the contagiousness of the copy. He also believes that there is a non-Platonic copying, and he calls this "mimetic desire."[45] There is an object—for example, a Louis Vuitton bag. Someone else—for example, Uma Thurman—desires this bag, indeed owns it, ownership and possession being the apotheosis but also the end of desire. Her apparent passion for this object, artfully conveyed by the advertisement, stimulates my own desire for it. I imitate her desire—it is contagiously transmitted to me; and now I feel that "I have to have it!" For Girard, the consequences of this triangulation of the object are considerable. Uma and I become rivals for the object. Through an act of displacement that has everything to do with the capitalist logic of the copy, it turns out that there is more than one Louis Vuitton bag. Thus, although my mimetic desire for the object has been stimulated by advertising and by fashion, the energy of desire, competition, and rivalry, which would otherwise lead to violence, will be channeled by market forces: by how much I am willing to pay for the bag, and by how many bags LV makes and how much the company is willing to sell them for. And LV has evolved a highly sophisticated modulation of mimetic desire: the company doesn't sell bags through other retailers; it doesn't offer discounts or other pricing incentives. What is most profound in Girard's work is the possibility that the object, in this case the Louis Vuitton bag, *exists as* the sum of mimetic desires focused on it, and it takes on its full meaning within an economy of displaced envy, jealousy, or, to use a key Buddhist term, attachment. But the contagious dissemination of those desires inevitably passes far beyond Vuitton's "official" production of the bag, for the displacement of this desire can be enacted not only with the official copies of the bag produced by the company, but by those other unofficial copies made by counterfeiters.

The Bonding of Louis Vuitton

In the absolute sense, from within the ocean of nonconceptual sameness there is no Louis Vuitton. In the relative sense, Louis Vuitton, the name, the brand, and the bags emerge through the processes of fabrication, bonding, and mimetic desire that I have described above, as does the vast cloud of possible bags, similar or not, fake or original. As LVMH and other branders know, the things of this world are impermanent, and it takes enormous amounts of work to keep a Louis Vuitton bag a Louis Vuitton bag, to prevent it from slowly sliding into the entropic morass of bags on sale at Printemps and Galeries Lafayette (neither of which sells Vuitton), all of them perfectly reasonable places to store a wallet or a cellphone, some of them even cautiously imitating the Vuitton monogram, replacing "LV" with a "G," reconfiguring the cute diamond-like icons. The Louis Vuitton bags made by LVMH assume their power and authority through complex processes of bonding, which hold together a particular impermanent structure known as a Louis Vuitton bag. And these processes are often available to varying degrees to other parties who can thus create what are called "copies" or "replicas" of Louis Vuitton bags.

I am not arguing that all Louis Vuitton bags are the same in the relative sense. Vuitton interests me because of the unusual degree of care which the company has taken to preserve and/or construct the authenticity of its brand: from the images of Scarlett Johansson clutching a monogram bag and bestowing her star-power on it through tactile mimesis, to the evidence on the LV website that real French people (or Basques) worked on the bag and that the mode of production is therefore of a different order from that of other bags whose production is outsourced; to the undeniable charm and glamor of Marc Jacobs, who oversaw the design of the bags, adding his New York City street-savvy style to the classic, European glamor of LVMH; to the pleasure of going to a real Louis Vuitton store, with a splendid design by a world-renowned architect, a historical display of Vuitton trunks from the beginning of the twentieth century, where your LV bag will be carefully wrapped and bagged for you. All of this "works" to produce the supplement of essence or origin—the company is profitable; and for those who want them, bags that can be legally designated as originating from the company of Louis Vuitton, *malletier*, can be acquired. And all of this carefully orchestrated fabrication and bonding of signs can at no point reveal any essence of Louis Vuitton in any Louis Vuitton bag.

In his essay "Economimesis," Jacques Derrida talks about how for Kant, the productions of the fine arts of poetry or painting are separated from the world of crafts because the latter has an obvious economic incentive attached to production, while the poet and painter produce apparently without regard to such economic considerations. Derrida shrewdly notes that this separation of poetic and painterly production from the "lower" world of the craftsman is itself economic, and part of an overall system of exchanges which Derrida calls "economimesis"—the inextricably mimetic quality of all formulations of economy.[46] Historically, this moment of separation of the realms of art and crafts is also the moment of the invention of copyright law.[47]

Louis Vuitton embodies many of the consequences of this separation, working the boundary between fine art and craftsmanship, but also, more importantly, between art and crafts and the commodity and mass production. The company associates its products with the world of fine arts in the hope that the aura of the unique art object will also be transferred to their bags and clothes, enhancing the atmosphere of craftsmanship, which may have existed in the time of Louis Vuitton himself, but which has surely been transformed in the age of global mass production and intellectual-property law. IP law paradoxically, and perhaps impossibly, demands that the discourse of essence, of original expression and uniqueness, be continually asserted in order for a monopoly on the right to mass-produce particular items to be maintained. Copying must be disavowed, aesthetically and legally, even as it supports the entire vast apparatus of production.

In this regard, the collaboration between Vuitton and Takashi

Murakami, with which I opened this chapter, takes on a special significance. Ever since the days of Andy Warhol, copying has played a powerful and explicit role in contemporary aesthetics; but what began as an attempt to establish a fundamental and subversive uniqueness at the heart of mass production—the Campbell Soup can as unique art object—has become part of the discourse of intellectual property, where "unique expression" has a particular legal meaning with profound economic consequences. At the same time, everybody now knows that the cheesiest mass-media image or mass-market commodity and the most rigorous abstraction by Barnett Newman or Jackson Pollock contain exactly the same degree of emptiness, and this recognition fuels the extralegal fascination with copying. Part of the brilliance of Murakami's work is its implicit and explicit recognition of all of this. The title of his 2008 show, "Copyright Murakami," cited—and packaged—it all.

Murakami has coined the term "superflat" to describe the absence of distinctions between high and low which could separate "art" from "craft" in Japanese cultural history. If we extend this line of thought, the Vuitton logo is as empty as the color fields in which Murakami situates it—it is "the same" or "superflat"—indeed, the term "superflat" resonates with the concept of emptiness, while subtly reifying it by transforming the metaphor of space or spaciousness often used in Mahayana discourses of emptiness into that of a plane or surface (which are already concepts and thus not truly empty). Nevertheless, the surface of the LV handbag and the exhibited canvas are also "similar" in this sense. And the fake bag sold on the street and the real bag sold in Vuitton's Paris store are also equally empty—and connected. Murakami and Vuitton play with this recognition of sameness, even as they defend their trademarks and copyright on their products. Indeed, it is necessary for their respective practices. There is no Louis Vuitton without the copies, the fakes that Vuitton continually distinguishes itself from through its various branding practices, which in turn generate further fakes. Murakami does not try to distinguish himself from either Louis Vuitton or the proliferation of copies around him, yet he protects the copyright on his work. He wants it both ways: the right to participate in the flow of interdependent, empty, groundless nonduality, and the right to claim this participation as exclusively belonging to himself. In the current legal, economic, and political regime, and in particular with intellectual-property laws which channel production into certain heavily overdetermined categories such as "art" and "branded product," we are all forced to engage this impossible ideological double bind. But no matter how "superflat" Murakami claims his paintings to be, they are in fact . . . *empty.*

2/Copia, or, The Abundant Style

Interviewer: How do you define folk music?
Bob Dylan: As a constitutional replay of mass production.

—Dylan interview, December 3, 1965, San Francisco, at 25′15″, *Classic Interviews, Volume 1*, www.dylannl.nl

All Praise to the Goddess Copia

The word "copy" comes to us from the Latin word "copia," meaning "abundance, plenty, multitude."[1] Copia was also the Roman goddess associated with abundance. Very little is known about this goddess, but she is mentioned in Ovid's *Metamorphoses* at the point where Achelous transforms himself into a bull in order to overcome Hercules, who responds by breaking off one of his horns. "But the naiads filled it with fruits and fragrant flowers, and sanctified it, and now my horn enriches the Goddess of Plenty."[2] Copia is depicted on a Roman coin with this horn of plenty, overflowing with the bounty of the earth, from which we get the word "cornucopia."

When we talk about copying today, when controversy around

copying occurs, these meanings of "copia"—coming to us from before the age of print, the age of mechanical reproduction, or the age of the computer—reassert themselves. Although we no longer associate copying with abundance, but link it rather with the theft or deterioration of an original, and thus a decrease, the phenomena we label "copies" and the activities we call "copying" still manifest this abundance and this increase. Copia as abundance continues to speak to us as a trace reverberating through the shifting historical meanings of the word "copy," and various practices of copying that are prevalent today still evoke the goddess, even if the practitioners no longer know the meaning of her name.

In his recent book *Free Culture,* Lawrence Lessig writes a manifesto for a free culture that seems strangely divergent from the practice of freedom as we know it on the Internet today. This divergence occurs because, when we use the term "free culture," we are doing more than merely trying to define a space in which certain creative uses of intellectual property are legitimated. The free culture that really interests us is the one described by a character in the remarkable science-fiction novel *Roadside Picnic,* by the Russian Communist-era writers Arkady and Boris Strugatsky: "Happiness for everybody! . . . Free! As much as you want! . . . Everybody come here! . . . There's enough for everybody! . . . Nobody will leave unsatisfied! . . . Free! . . . Happiness! . . . Free!"[3] What appears to be on offer on the Internet, what fuels its imaginal space, is the utopia of an infinite amount of stuff, material or not, all to be had for the sharing, downloading, and enjoying. For free. And this too is Copia's domain, which can still be accessed today through "copying."

In the Western imagination, such moments of being overwhelmed with an infinite amount of desirable stuff are epitomized by feasts, with tables stacked to the rafters with tasty foods—by festivals in which diverse kinds of sensual pleasures come together in a mass of bodies and sensory stimuli. We think of treasure caves where gold, jewels, precious objects are hoarded in vast mounds, of genies who grant wishes. We think of marketplaces in which the goods of the world are spread out; of department stores like Harrods in London, Bloomingdale's in New York, or Galeries Lafayette in Paris, and shopping malls like the Eaton Center in Toronto, where every imaginable consumer item is on display.

If you want any part of these last fantasies, you're going to have to pay for them. You can enjoy them as spectacle, going window shopping, as my mother and father used to do in suburban London when I was growing up. But if you want more intimate enjoyment, you need money. Or a strange twist of fate, like the one that occurred on July 13, 1977, when the power grid went out in New York City, leading to widespread looting in poorer neighborhoods such as Harlem and the Bronx. It is that day which is credited in *Yes Yes Y'All,* a recent oral history of hip-hop, as being the moment of hip-hop's tipping point, where the technologies required for MC-ing and DJ-ing (turntables, microphones, and speakers), formerly available only to a small number of crews, were suddenly in the hands of just about anyone who wanted them. This free access facilitated hip-hop's full emergence as a culture. Or one might consider the day in fall 1999 when Shawn Fanning released the first version of Napster, facilitating an explosion of filesharing which peaked in February 2001, when 1.6 million users had access to free digital copies of millions of audio recordings.

We have a word for such activities: "stealing." And stealing is punishable by law. Don't the store owners, musicians, writers, and software programmers whose work is suddenly made available in these free-for-alls deserve to be compensated? How would you like it if someone came and stole your stuff, or—to return to the theme of my previous chapter—made copies of all your work and sold them or distributed them for free without your permission? In terms of the current legal, economic, and social regime, these questions are all valid. But below the surface of contemporary consumer culture, there is a collective dream of free access to an infinity of things. It is one of the principal themes which advertising manipulates, except that "free access" has been replaced by the promise of access via the purchase of a product—say, a soda or a pair of sneakers. The crises around property that are marked by the blackout riots in New York, or by digital filesharing, tell us that radical shifts are taking place in these different regimes. And the word "copy," a ubiquitous but poorly understood word, is playing an active role in these shifts. This word cannot be restricted to the particular set of definitions that we currently give it—any more than the appropriations of the 1977 blackout or of digital filesharing, so productive for the cultures and communities they helped to mobilize, can simply be dismissed as a crime.

The Origins of Copia

Who was Copia? Aside from the lines by Ovid quoted above, she appears to be a thoroughly obscure figure, usually explained away as a product of the Roman predilection for turning abstract principles, particularly those associated with personal gain, into deities. She barely appears in even the most comprehensive resources on the classical world. But the word "copia" was in common use, meaning "abundant power," "wealth," "riches," "abundance," "fullness," "multitude."[4] If these senses of the word are still familiar to us in the word "copious," others are more unusual: "copia" had a military meaning as "a body of men," and a general meaning of "storehouse," "a set of resources at one's disposal," "the means, possibility, or opportunity of doing something."[5] The word "copia" is derived from "cops" ("abundance"), and "cops" is derived from "ops" and either "con" or "co." This is a matter of some significance, since it links "copia" to a rather more well-known goddess Ops, who was also a goddess of abundance, associated with the harvest, and with another harvest deity, Consus, who was the protector of grains and of the storehouses in which the harvest was kept.[6]

We pass further into the labyrinth of Roman mythological etymologies at our peril; but in tracing the origins of "copia," we find a god/goddess pairing relating both to the overflowing bounty of the harvest and to its storage for use. And copia itself contains this dual sense: abundance, but also the deployment of abundance. And in this double meaning, one can already discern some of the qualities that will come to the fore in the word "copying"—the copy as an object that is inherently multiple, that is more than one, that is a copy of something, and thus part of an excess or abundance, of a *more*. And at the same time, the copy is part of a storehouse, an object created or appropriated in order to be an object of use, made part of a store that is available; and as a part of a store, something that is counted or measured, named and/or labeled, owned, and no longer freely existing for itself.

The word "copia" appears to have emerged in Rome when Ops, the harvest goddess, and therefore a goddess of the countryside, was transplanted to the city, where she was honored with a temple on the Capitoline Hill, one of a series of deities who functioned as personifications of virtues or abstract qualities and whom Cicero talks about. Thus, Ops became a more general goddess of prosperity, associated with the protection of the city. She became associated with copia (abundance in general) as well as with *auxilium* (a unit of troops). At the origin of copying, then, we find . . . a copy! For Copia was already a copy of a goddess, an appropriation of Ops made in the transplanting of the nature goddess to the city, manifested in a culture where phenomena that were easily related to what we today call "copying" abounded—from the appropriation of Greek and other cultural models by Roman culture, to the invention of substances such as concrete which are so useful in producing replicas, duplicates, multiples, "copies"; to the mass production and circulation of multiples of various products such as oil and wine in generic amphorae (vessels); and, more broadly, to the imperial implementation of a generic "Roman culture" across the empire.[7]

Copia was clearly a goddess of economy in ancient Rome; and according to Jacques Derrida, mimesis will in every case be a matter of economy.[8] Every copy, every act of exchange, presupposes the establishing of an equivalence between *a* and *b*, the assumption that they are like or equal to each other in some way. There are different kinds of economies, all of which manage or appropriate mimetic energies. There's the *sacrificial economy,* which Girard sees as being the predominant one: Copia, as a goddess of abundance to whom sacrifices were made, would be part of such an economy. There's the *capitalist economy,* where everything is made equivalent through exchange value and money—thus the Louis Vuitton bag, whose identity is established by being bought in an official Louis Vuitton store at the price set by the company. There's the *gift economy,* where things are exchanged and given meaning through complex systems of reciprocity in which an excess is always part of the process of gift giving and taking, so that the copy is never "the same" and always part of a dynamic, shifting abundance.

The sacrificial economy, crucial to Rome and to the emergence of Christianity within the Roman Empire, today takes the form of the legally encoded economy in which certain people are scapegoated and punished for making and exchanging the same copies that everyone else is making and exchanging. The word "copy" appears today at all those locations where the dominant capitalist economic structure stutters and stumbles. Copying and the crises that surround it today are the sign of an economic hesitation, the manifestation of traces of some other economy, future, present, or past. New technologies such as the computer or the Internet open up issues of economization ("monetization" being only one particular kind of economization), and a variety of economic trajectories that are not easily assimilated to the current structure. "Copies" appear and are labeled as such out of the vast plenitude of mimetically appearing objects, at moments when those objects cannot be fit into the social/political/economic system as it evolves. Thus, they appear as the markers of the danger of an excess or abundance that needs to be controlled.

The Abundant Style

The word "copia" was also associated in ancient Rome with rhetoric. *Copia verborum* ("abundance of words") referred to the copiousness of language, the storehouse of words and rhetorical techniques at the disposal of one skilled in the art of rhetoric. From classical times to the Renaissance, there existed manuals of rhetoric that advised people how to speak and write. These manuals were the basis of scholarship and public discourse. They were also concerned with imitation, since their subject matter was considered to be not something original, but the continuation and repetition of a tradition that had begun with the ancient Greeks. These manuals were not designed to instruct people to imitate or copy per se—although Erasmus, author of *In Praise of Folly*, a book whose title I cite, copy, or steal in my own book here, wrote a celebrated rhetoric manual called *On Copia of Words and Ideas* (1512).

"Copia," according to a contemporary of Erasmus, meant the "faculty of varying the same expression or thought in many ways by means of different forms of speech and a variety of figures and argument."[9] The three components of rhetoric, *inventio* (the selection of matter or elements), *dispositio* (the arrangement of those elements), and *elocutio* (the style of presentation), did not include imitation per se, but it was understood that the practice of imitation was fundamental to rhetoric. This was a matter of some concern—the Roman rhetorician Quintilian, for example, stressed that good rhetoric could not *just* be imitation.[10] Thus, we can see a gap opening up between mimesis and copia, between copying understood as a crude act of thoughtless repetition (Quintilian's main objection to a speech that is solely imitation is that it does not charm the listener) and copying as the many possibilities for variation within the act of repetition.

The translators of Erasmus, perhaps squeamish about using such a degraded word, refuse to translate the Latin word "copia" as "copy," but in medieval and Renaissance England, "copy" (or "copie") was the standard translation of "copia," and had the meanings of abundance, multiplicity, which are still contained in the word "copious" today. While "copy" was used to denote a duplicate of a text as early as the fourteenth century, the more general meaning of "something made or formed, or regarded as made or formed, in imitation of something else" did not emerge until the end of the sixteenth century.[11] It was also around this time that "copia," which has an affirmative sense of resources, power, or plenty, started to take on a pejorative meaning: the copy as a degraded version of an original.

The reasons for this shift are connected with the emergence of the printing press, the book, and other technologies of mass production, and the process by which sets of legal controls and guarantees concerning the right to make and sell copies came into being. While copyright law itself did not emerge until 1709 in England with the Statute of Anne, patents were granted in Italy and England as early as the fifteenth century, and patents controlling the "rights in copies" of books can be dated to 1563 in England.[12] "Copye," in the sixteenth and seventeenth centuries, had an ambiguous meaning when used by publishers, since it referred both to the text which the publisher had the right to publish (the "original"), and to those copies of the original "copy" that were made by authorized publishers as well as by unauthorized parties. It appears that the concept of the original or authentic text, as something separate from the copies made from this original, was absent at this time, and only emerged in the eighteenth century with the evolution of Romantic aesthetics.[13] Thus, English poet Edward Young wrote in 1759 that "An *Original* may be said to be of a *vegetable* nature; it rises spontaneously from the vital root of Genius; it *grows,* it is not *made: Imitations* are often a sort of *Manufacture* wrought up by those *Mechanics, Art,* and *Labour,* out of pre-existent materials not their own."[14] After describing the Museums of Copies that existed in Paris in the mid-nineteenth century, Rosalind Krauss notes that in nineteenth-century France "the copy served as the ground for the development of an increasingly organized and codified sign or seme of spontaneity." In other words, the concept of an original could not exist without that of a copy, and, in practice, "originality" was not an objective fact but a historically specific style of presentation—a recognizable roughness, spontaneity, or naturalness, for example.[15] And these words would undergo a further shift of meaning after World War II in the work of John Cage and the Fluxus group; William S. Burroughs, Brion Gysin, and others associated with the Beats; and Andy Warhol and various Pop artists—all of whom argued that the copy was more original than the original, precisely because it made explicit its own dependence on other things, signs, or matters.[16]

Folk Cultures and the Death of Copia?

In his study of copia in the Renaissance, Terence Cave observes that although writers such as Erasmus and the gloriously copious Rabelais were fascinated by copia, they actually believed that they were living in an age in which abundance itself was dying, declining into mere verbal profuseness.[17] In the seventeenth century, rhetoric as a self-conscious practice built on classical models faded in the face of the new Cartesian emphasis on method and the growing importance of scientific measurement, which called for a modest, more precise style of writing. While the Industrial Revolution brought with it new forms of abundance (see Chapter 5), these were carefully locked into the new logic of the marketplace, and new economic and political frameworks.

But did copia in fact die? The belief that copia as abundance is today a mere utopian fantasy is a crucial ideological determinant of the history I have set out above—the process by which the celebration of abundance as a part of nature and cosmos retreated into language and literature during the Renaissance, where Copia's power became symbolic or "rhetorical," in the more limited, modern meaning of the word.

Cave's literary view of copia is certainly open to question. Mikhail Bakhtin, in his famous study of Rabelais, argues that the grotesque excesses and humor so evident in Rabelais' work *did* reflect the worldview of medieval and Renaissance peasantry.[18] In fact, "copying" is found everywhere in the festivals that Rabelais describes: the extensive use of satire and parody functioned as humorous imitations and inversions of the feudal, church-dominated medieval world. Bakhtin opposes the seriousness, properness, and completeness of official dogma with the ever-changing, decaying and regenerating, incomplete cycles of the natural world which the medieval peasantry celebrated. The banal, frozen image of the cornucopia came up against the more scandalous but accurate image of abundant nature as an endless, changing profusion of forms produced and reabsorbed. Although it is difficult to recognize the mass of "degraded" (to use Bakhtin's word) grotesque forms in Rabelais as copies, this "degradation" as a sign of the incomplete, ever-changing multiplicity of things and beings remains part of our discourses of copying today—but the affirmative, political quality of the "degradation" is now hidden or forgotten.

Surprisingly, Bakhtin, who speaks so eloquently about medieval and Renaissance peasant culture, did not believe that the grotesque culture of the medieval marketplace, with its celebration of an exuberant grotesque abundance, had survived into modern times, except in odd traces. It is difficult for us to imagine copia today outside the laws of the marketplace, which label, measure, and define copia and abundance so that they become almost unthinkable outside the monetary system and legally or scientifically defined entities. But as shown by the examples of Napster, the 1997 blackout in Harlem, or the documentable persistence of carnival and festival (particularly in the global South), copia as abundance persists; and at brief but decisive historical moments, this idea surges forth as a transformed, vital reality.

The German philosopher Johann Herder coined the term *Volkslied* ("folk song") in the eighteenth century and produced a two-volume collection of folk song lyrics from around the world. But there have always been folk cultures, usually existing in the shadow of kings, churches, rulers of various kinds. The peasantry, out of necessity, out of the fact that they owned little or nothing, found "unofficial" ways of making, distributing, and sharing things—like songs, for example, or recipes or spells. They developed particular collective techniques for producing these things—appropriating, cutting and pasting, transforming whatever came to hand, using what anthropologist Claude Lévi-Strauss called "the science of the concrete." Then industrialization came along, and with it new kinds of "official" distribution networks—the capitalist marketplace, copyright and intellectual-property law, and the Romantic cult of the individual artist, who at the same time sold his or her work in the marketplace like any other worker. In the nineteenth century, European folk cultures apparently disappeared as autonomous entities. Elements of the political right appropriated and represented them as reified kitsch symbols of the nation-state. Such reifications were rightly seen as fascist manipulations by the left, who nevertheless also embraced industrialization and the transformation of the peasantry into the proletariat. Marx's proletariat were the slaves of a particular regime of copying—industrial capitalism—but the "lumpenproletariat" or "subproletariat," which included "swindlers, confidence tricksters, brothel-keepers, rag-and-bone merchants, beggars, and other flotsam of society," were already engaged in a broad and autonomous set of practices of copying that were continuous with those of folk cultures. If Mike Davis in his recent book *Planet of Slums* is to be believed, people in this situation now constitute a majority of the world's citizens; and in the 1960s Frantz Fanon, the Black Panthers, and the Young Lords all recognized (briefly) their revolutionary potential.

Today, the activities of the urban poor, not to mention the activities of other groups which cannot be fully integrated into the market system, go by a variety of names: "subcultures," "subaltern cultures," "the governed," "the masses." All could equally be termed "folk cultures," defined as semi-autonomous collectives and groups gathered around certain practices and values. These practices might include: socially oriented musics, from hip-hop and its global variants to dancehall, Celtic folk, metal, and goth; the activities of avant-garde groups ranging from Dada and Futurism through mail art, happenings, and punk to present-day fanzine, website, or performance-producing collectives; popular literary forms that respond directly to the industrial world, from science fiction to Romance to pornography to various kinds of street pamphlets and literature; new hybrid religious forms, drug cultures, and sexual subcultures; street gangs and other urban collectives, such as bikers and flash mobs; ad hoc women's groups trading material and aesthetic practices that transform everyday objects, such as quilting bees, recipes, medicines, textiles, garments; computer-based cultures such as hackers, filesharers, bloggers, and other posters and lurkers; dumpster divers, scavengers, denizens of Salvation Army and Value Village thrift stores, habitués of sidewalk, church, and garage sales; practitioners of martial arts, dances, rituals of many kinds. There is no obvious equivalence among these groups and activities, which span the reaches of the industrial world from its margins to its absent center. All are transnational forms which develop locally in unpredictable ways. This is "the multitude," as Michael Hardt and Antonio Negri recently characterized it, or, in Ernesto Laclau's formulation, "the people"—in each case a mass of individuals characterized by their heterogeneity, a quality which also renders them subordinate. These groups are all associated with practices of copying which render them inauthentic, abject, in different ways, but which are also the mark of a certain autonomy. Gayatri Spivak notes "the lack of communication within and among the immense heterogeneity of the subaltern cultures of the world."[19] I do not wish to elide the obvious differences between a downloader of films sitting in a dorm room at a North American college and a vendor of shopping bags made out of used sacking in a market in a suburb of a city of the global South. But what if it is precisely practices of copying, the affirmation of copia, a particular attitude toward mimesis, that constitutes what these diverse groups have in common—and which makes them illegal, illegitimate, or marginal?

Playlists and Mixtapes

Let's consider specific examples of copia as it is found in contemporary culture. I am not going to focus on the purity or lack thereof of the folk forms which I discuss; the difficulty of recognizing copia today is related to the problems we have seeing beyond capitalist cultural forms. The iTunes playlist, with its MP3s purchased from the iTunes Store or downloaded from other sources, embodies the contradictory aspects of copia in contemporary society: the ability to manifest, exchange, and share a vast abundance of "copies" of sound recordings, limited only by storage capacity and access to the appropriate computer technologies; also, the framing of this abundance through a database acting as a storehouse and presenting it as a set of discrete units; the plenitude of sound, turned into code, which in turn is transformed into a named file stored on a hard drive. And the commodification of this system via iTunes, with its complex systems of digital property rights management, where the abundance of sound-worlds that forms a part of copia is presented as a variety of commodities available for purchase within a highly organized legal-economic system.

iTunes participates in the now-standard contemporary capitalist practice of taking a folk or "subcultural" form, usually involving some aspect of collective folk play, and commodifying or recommodifying it (since most subcultural practices involve the appropriation of commodity forms). This doesn't necessarily make iTunes uninteresting. It's striking how many of the "utopian" aspects of copia are present within iTunes: it is free, and freely available; it plays both files purchased and files exchanged within the gift economy of peer-to-peer and other networks. What Apple has done is to insert commodifiers within this abundance: the iTunes Store as a legal and financial transaction-based way of gaining access to the abundance of the world of sound; the iPod and the Apple computer as technologies for storing and manifesting this vast world of sound; digital rights management programming that controls how many copies you can make of files purchased at the iTunes Store.

It will be easier to recognize the presence of copia if we consider a recently outmoded production of an industrial folk culture that is related to iTunes: the mixtape. What exactly is a mixtape? Usually, a ninety-minute miniature reel-to-reel of magnetic tape, with a selection of songs recorded from the radio or vinyl LPs or other cassettes, wrapped in a little clear plastic box with a cardboard insert with the names and producers of the tracks listed in felt-tip pen and maybe a photo clipped from a magazine for a cover. Most people would call a mixtape a "copy" in the derisory sense of an inauthentic repetition or recreation of an original. It is only with the recent demise of the mixtape in favor of the mix-CD or iTunes playlist that we are able to see the extraordinary kinds of creativity that actually went into making mixtapes. As Dean Wareham has written in the book *Mix Tape: The Art of Cassette Culture,* edited by Thurston Moore:

> It takes time and effort to put a mix tape together. The time spent implies an emotional connection with the recipient. It might be a desire to go to bed, or to share ideas. The message of the tape might be: *I love you. I think about you all the time. Listen to how I feel about you.* Or, maybe: *I love me. I am a tasteful person who listens to tasty tunes. This tape tells you all about me.* There is something narcissistic about making someone a tape, and the act of giving the tape puts the recipient in our debt somewhat. Like all gifts, the mix tape comes with strings attached.[20]

Wareham and others in the book capture beautifully how intimate the act of making a copy can be, how emotionally charged the act is, and what a complicated repertoire of gestures the mixtape maker has available in accessing the variety that is essential to copia, and straight out of a classical rhetoric manual: *inventio* (the selection of tunes to be played), *dispositio* (the ordering or sequencing of them), and *elocutio* (the cuts and edits made, but also the loving care put into the handwritten cover, the decoration of the cassette), all deployed in order to charm the recipient of the tape.

The course I teach is always energetic when we come to discussing music—partly, no doubt, because students don't have to read anything much in order to talk about music. So in recent classes there was considerable enthusiasm for the topic of mixtapes. We talked about the difference between mixtapes and mix-CDs, and everyone agreed that mixtapes had a handcrafted quality that was vastly superior to the impersonal quality of mix-CDs, which are produced automatically by a computer program after you make a playlist. Both tapes and CDs are industrial-era products (well, CDs are post-industrial), so it's interesting to see the differences: the importance placed on the magic of the hand with cassette tapes, the power of tactility to confer magical powers on the copy. In this case, it is the hand on the Pause button, coordinating with a turntable or a CD player or the radio, and the magical trace of this hand adding something personal and powerful to the recording—whereas the CD's only trace of the hand is that of the movement of the mouse in arranging names on a playlist and clicking on a "button" on the screen to burn a CD. Digital recordings present real problems in terms of mimesis, because the tactility that is key to mimetic magic is lost. But when I asked the students whether they'd rather receive a mix-CD or a mixtape, everyone said they'd rather have a CD, because the digital likeness of the copy is more accurate than that of the cassette.

So does this mean that likeness trumps tactility in the production of the copy? There was considerable confusion on this point. On the one hand, the students recognized the work that went into the tape and valued it; yet at the same time, they valued the CD's quality of reproduction more. It is as though the abundance of copia had undergone a split into the variousness of the individualized tape, and the more numerical, quantitative, and "accurate" abundance of the digital file. One of my students told a beautiful story. He had made a mixtape of a bunch of MP3s for his girlfriend, and they'd transferred it to a cassette tape so that they could listen to it while driving around in the car. Time had gone by, maybe the relationship had fallen apart, and after reading the book *Mix Tape* for the course, the student dug around and found that cassette. He set about ripping the cassette onto his computer to make a CD, which he planned to give to his girlfriend. This made me think that maybe what produced value wasn't just the labor of manually clicking the Pause button on and off for all those vinyl or CD tracks, or the economic value of giving someone $100 worth of tunes, or the mimetic value of "true likeness," both of the tracks and the giver, or that mimetic tactile magic of the hand on the Pause button, encoded onto magnetic tape. Perhaps what created worth was a kind of *transformational* value. In Locke's classic definition of property, it is one's labor that allows one to appropriate something from nature and call it one's own. This is an eminently mimetic theory, in accord with the tactile, contagious quality of mimesis that we examined in the previous chapter. But there is no reason that this touch, which leaves a trace of the student on the object, should imply private property. Rather, it is a moment of meaning, of exchange, of contact.

A Show by Bert Jansch

> Folk songs are evasive—the truth about life, and life is more or less a lie, but then again that's exactly the way we want it to be. We wouldn't be comfortable with it any other way. A folk song has over a thousand faces and you must meet them all if you want to play this stuff. A folk song might vary in meaning and it might not appear the same from one moment to the next. It depends on who's playing and who's listening.
>
> —Bob Dylan, *Chronicles* (2004)

I recently attended a show by 1960s British folksinger and guitarist Bert Jansch, who is considered a master of folk music. Like most folk culture, folk music is built around traditions—ways of doing things, particular forms such as songs that are passed down from performer to performer. There are rules about what can be done and what can't: when Bob Dylan brought electric guitars to Newport in 1965, people were outraged. If the repetition of forms is a necessary part of folk culture, how do individual performers avoid boring their audiences? Jansch's performance was exquisite, and even though much of his material was traditional, it became clear that he was using a number of devices or methods to bring his particular performance to life in front of that particular audience.

The first thing to note is that Jansch appeared to choose his songs out of a seemingly encyclopedic number that he knows. A copious number. The act of choosing, emerging out of dialogue with the audience already brings spontaneity, living presence, to what could be generic or repetitive. When Jansch introduced a song, he situated it within the tradition of singers that he comes from, telling a story about who taught it to him, or the exact moment he first heard it. "X taught me this song, when I was managing a pub in Glasgow." Thus, although the object being copied is generic, the moment and circumstances of Jansch's encounter with the object are unique. Not only does each moment of copying become unique, since it is connected to a series of unique moments in which the various singers encountered the song, but the object being copied—say, a traditional song like "Reynardine"—itself dissolves into a million slightly different versions, all with their own history and trajectory. I was struck by this recently when I tried to find the lyrics to "O Bury Me Not on the Lone Prairie," and could not find a version of the song that was identical to the one I'd heard La Monte Young sing. What I found were many slightly different variations—all recognizably the same song, but all slightly different. Just as Dylan says in the quote at the head of this section.

If the moment that a singer learned or first heard a song is important, setting forth the moment that the song was given as a gift to the singer, then so is the moment in which the singer presents the song to an audience. The act of storytelling establishes the uniqueness of both the history of the song and, more important, the uniqueness of the occasion on which the song is being sung. The song is being presented to a particular audience; and just as I introduce myself when I arrive at a party or a dinner, the song too is introduced. This introduction presents the song to the audience, offers it as a gift, and establishes a reciprocal obligation in the audience: to buy tickets, to listen, to respond. This tradition can take many forms—for example, I recently saw Syrian singer Omar Souleyman perform. Souleyman mostly performs at weddings in Syria, and works with a poet who appears onstage with him, and whose job it is to compose verses in honor of particular guests who request them and pay him for them. He does this with the help of a laptop at the side of the stage, and whispers the words to Souleyman who then sings them. The presentation of the song as a gift transforms its status as a copy. In this sense, the sharing of MP3s through a peer-to-peer (P2P) network, a mixtape, or a podcast, all of which are acts of participation in gift economies, itself changes the nature of the copies that are being passed around. P2P networks have gone so far as to encode the reciprocity of gift giving through the monitoring of upload/download ratios.

I was also struck by the ways in which Jansch ended songs. Walter Benjamin writes about the importance of ornament as the set of possible ways in which a perception of a particular form might emerge out of a pattern or a weave.[21] Although the endings to Jansch's songs are highly appropriate, they have seemingly little to do with the songs themselves. They are like the last flick of a pen signing a signature, trajectories by which the writer or performer trails off, leaves, stops. Embellishment, or ornament, is itself a vast topic, from the rhetorical manuals of medieval Europe, to the gamaks of Indian classical music, to the stylized gestures of Japanese Noh theater and Indian Bharat Natyam dance—and Jansch embellishes each song in his own way. The most minute gestures change the mood of the song; and among those most passionate about folk cultures, an aesthetic of subtlety prevails, devoted to the most minute yet powerful gestures—an art of appropriateness, of honoring that which makes an object, a form, what it is, but an art which at the same time can open it up, reveal it, adorn it in different ways. Embellishment, in this sense, is not just the ornamentation of a preexisting object, but an integral part of how the "copied" object manifests itself in multiple, diverse ways. The topic is a complex one—for example, the concept of *rasa,* enormously important in traditional Indian aesthetics, can be translated as "essence" or "juice." This essence is evoked by a pattern of gestures, musical notes, words, or acts which are not symbolic of an idea but actually manifest it. In this sense, rather than speak of the ubiquity of essencelessness, we might more accurately speak in comparative terms, of different ecologies of essence and essencelessness. In the case of *rasa,* ornament *is* essence.[22]

The Cloud of Copia

How does the folk view of copia-as-abundance relate to the philosophy of copying based on the inherent emptiness of all phenomena that I set out in the previous chapter? If there is abundance, doesn't this prove that there is some *thing* that is abundant, some thing of which there is more than one, and that that thing necessarily *is?* But the multiplicity of copia is predicated on the emptiness of that which is, even from a relative point of view, never "the same," in the Platonic sense. For the sake of specificity, I will talk principally about Buddhist theories of emptiness here, but we should retain the thought that different figurations and structurings of what is called "emptiness" appear in different cultural and philosophical traditions around the world.

The connection between emptiness and abundance or multiplicity appears in a number of Buddhist texts, including the *Avatamsaka,* or *Flower Garland Sutra,* a core Indian Mahayana sutra that became the foundational text for the Hua Yen school of Buddhism that flourished in China during the T'ang Dynasty. The *Avatamsaka* is notable because it is said to be the only sutra which describes the "worldview” of the Buddha while he was in the state of enlightenment he attained under the Bodhi tree in Bodhgaya—an infinite series of infinitely connected worlds or Buddha fields that scale from atoms to universes, each of them composed in relation to the vast multiplicity of all the other worlds, described in a series of psychedelic visions that add up to fifty books and fifteen hundred pages in the English translation.[23] The central image of the sutra is that of the Net of Indra, which is an image of this multiplicity of universes, every node of which is a jewel interlaced with every other jewel and reflecting every other jewel, while at the same time completely lacking self-existence.

> Then Universally Good also said to the assembly, “In the land masses of this ocean of worlds are seas of fragrant waters, as numerous as atoms in unspeakably many Buddha-fields. All beautiful jewels adorn the floors of those seas; gems of exquisite fragrances adorn their shores. They are meshed with luminous diamonds. Their fragrant waters shine with the colors of all jewels. Flowers of all kinds of gems swirl on their surfaces. Sandalwood powder settles on the bottom of the seas. They emanate the sounds of Buddhas’ speech. They radiate jewellike light. . . . There are unspeakable hundreds of thousands of billions of trillions of banners of ten precious elements, banners of belled gauze of raiments of all jewels, as many as sand grains in the Ganges river, jewel flower palaces of boundless forms, as many as sand grains in the Ganges river, a hundred thousand billion trillion lotus castles of ten precious substances, forests of jewel trees as many as atoms in four continents, networks of flaming jewels, as many sandalwood perfumes as grains of sand in the Ganges, and jewels of blazing radiance emitting the sounds of Buddhas’ speech; unspeakable hundreds of thousands of billions of trillions of walls made of all jewels surround all of them, adorning everywhere.”[24]

This baroque description, which itself is repeated with variations throughout the text, is a figuration of the relative world as a vast matrix of traces—infinite multiplicity manifesting a dynamic sameness beyond all concepts. In this multiplicity, there are certain words, modalities that have a special affinity with nonduality: fragrance, sound, jewels, color, light. Words like "adornment," "ornament," and "emanation" are part of a rhetoric of embodiment or manifestation which negotiates the paradox of how to figure or describe something without essence.

There is a precedent for such rhetorics in Buddhism, where the question arises how, if the Buddha has achieved enlightenment beyond all concepts, he can communicate the path to enlightenment to beings who remain in this relative world. "The Buddha" is often described as a multiplicity, with three bodies (Sanskrit: *trikaya*) manifesting according to different ontological and epistemological frameworks, along with 84,000 different teachings, each appropriate for a particular situation or being. Many Buddhist practices involve multiplication and repetition on a vast scale—for example, the repetition of a mantra. This could be done mentally. But it could also be done as an act of writing, as with the *mani* stones (rocks with the mantra "Om mani padme hum" carved onto them) which cover Tibetan sacred sites. Or with the printing of mantras on pieces of paper which are then inserted into statues of deities or other sacred forms. Or the printing of mantras onto prayer flags, which are hung from mountaintops and said to sow the wind with the prayers inscribed on them. Indeed, the oldest dated printed book known is a copy of the *Diamond Sutra* from 868 A.D. found at Dunhuang in Western China.[25] In the realm of sculpture and the visual image, anyone who has visited a Tibetan Buddhist temple will have been struck by the proliferation of images. At some famous temples such as Samye, there are rooms containing thousands of identical statues of a particular deity, made as an act of devotion to accumulate merit, but which focus the mind through the repetition of the image. And then there are *tsa tsa,* small molded-clay figurines made in vast quantities, again as an act of devotion, and often placed inside larger statues. The evolution of book printing in both India and China was connected with the merit to be gained from producing multiples of the Buddhist scriptures, and a mechanical or even mathematical relationship was established between quantities of books and quantity of merit—thus the invention of mechanical printing (in China) was thought to facilitate a rapid and efficient accumulation of merit.[26] Such technologies of spiritual repetition continue to be used—for example, the "Buddha machines" made in China today, transistor-radio-sized objects which contain recordings of particular mantras and tiny speakers, and which, when switched on, repeat the mantra until the batteries run out.[27] In each case, the multiplication of nearly identical images is understood not as the degradation of an original, but the invocation of an impermanent, provisional form with the goal of training the mind to recognize its own true nature.[28] The *Diamond Sutra,* key to the development of Buddhism in China, formalizes the relationship between copying and enlightenment as follows: "If there be one who hears this scripture and believes it unfalteringly, his merit will be greater than this. How much the more is this the case for one who copies, keeps, and recites the scripture, explaining it to others."[29]

In economic terms, the Net of Indra can be understood as a vision of what Georges Bataille called "general economy" (as opposed to restricted economy). Bataille was interested not only in the limited exchanges that are termed "economic" in particular societies, whether financial transactions or gift exchanges, but the total circulation of everything in the universe, from sunlight, to organic and inorganic matter, to planets. Economist E. F. Schumacher proposed the idea of a Buddhist economics.[30] By this, Schumacher basically meant an ethical economics. But this idea could be taken much further in the direction of Bataillean general economy. Buddhist practice and philosophy is always involved in the management, structuring, and directing of mimetic energy and the exchanges and equivalences by which it is organized. For the Theravadin schools, this means the refusal of all exchange, leading to nirvana as a state of extinguished desire. For the Mahayana schools, it suggests a total engagement in the vast network of dependently originated, relative existence in order to reveal the enlightened aspect which is immanent to it. The Net of Indra is the most complex figuration of economimetic exchange, for every jewel in the Net reflects every other monad-like component of relative existence, in the form of a chaotic flux.

The question of how a particular form arises out of a groundless, infinite, "excessive" multiplicity is a topic of considerable dispute in contemporary philosophy, notably in Alain Badiou's recent critique of Gilles Deleuze concerning his theory of the event. Their differences are about mimesis, since what is called "mimesis" is precisely the way a particular form arises out of a groundless, infinite excessive multiplicity. Both are concerned with understanding a phenomenon that is highly relevant to folk cultures—namely, those events where the vastness of a chain of multiplicities becomes apparent.[31] Such events obviously have potentially profound political consequences. Folk cultures are intensely interested in the moment of actualization, and the skill and knowledge that make the production of an event possible, and the conjuring of abundance in the form of a multiplicity of similar objects through particular mimetic practices which we might well call "rhetorics." But the rhetoric of a folk musician like Bert Jansch or of mixtape makers is not a matter of adding an ornament to a preexisting object that remains essentially the same but for certain bells and whistles. The rhetorics of copia are the modes in which things are made to appear with no prior or more solid ground than knowledge of the practices that lead to their production or revelation. Again, this is not merely rhetoric as discourse or social construction, but rhetoric as a mode of co-participation with materiality and other forces in the fabrication of objects and events which, along with the "subjects" that participate in their production, are all revealed as jewels in the Net of Indra during the festival, during the party, in all the various situations in which folk cultures do what they do. Certain cultural forms are more copious than others: music, in particular, appears and disappears fleetingly, conjures the immanent vastness of the Net, constellates into infinite sonic chains, precipitates collective joy, is eminently portable, and resists being turned into a thing or property—which is why folk cultures have such love for it.

Languages of the Marketplace

Walking through East London's Brick Lane recently, I was struck by the clustering of Bengali restaurants, most of which are serving almost identical food. It reminded me of the clustering of cloth vendors in a particular district of Chennai, or the Garment District in Manhattan. Or the markets of North Africa, such as Marrakech, where spice merchants, leather tanners, butchers, fruit vendors, and others all have their own areas. I grew up in suburban West London with the model of the High Street, where there are one or two examples of each type of shop in any neighborhood. Each shop establishes its uniqueness by the fact that it sells a different category of items or services, and has a different outward appearance (a different window display). The clustering of stores selling the same service in a particular neighborhood appears to me as a form of copying: the stores are "the same," they sell "the same thing." And the question arises in my head: How do I choose between the many stores around me that all look identical to me, that all look like copies of each other, and that I am unable to differentiate from one another?

The stores, however, have developed their own arts of copia, which will be familiar to anyone who has been in a North African souk or a Parisian street market. First of all, they devote great attention to visual embellishment and arrangement. The Bengali restaurants are festively lit and appear to be competing to produce the most splendorous storefronts and interiors using sparkling arrays of lights, neon signs, fabulous names. Although such accessories are not to be found in a souk, a strong visual aesthetic prevails there too: sacks of goods are displayed in arrangements of bright colors, and identical objects are piled up in glistening pyramids—an art of permutation and combination that plays at arranging the same in multiple ways.

Second, an art of charm or seduction is pervasive. Outside most of the Bengali restaurants are people whose job is to entice or persuade you to enter—much as the vendors in a souk call out to you to try their goods. In British street markets too, the fruit vendors try to outdo each other by shouting out the bargains that they are selling, extolling the virtues of their fruits and vegetables, flirting with potential customers. The attitude is playful but calls attention to the "variety" on display in a place where, because of tradition, the offerings are, from a certain point of view, "the same." Quantities are never exact, always a little over, or the price a little less. "A courgette for the baby!" said a vegetable seller at the market in Belleville, Paris, one morning, throwing in a little gift. Both buyer and seller scrutinize tomatoes with the expert eye of a Sotheby's art appraiser, alert to the subtlest signs of excellence or neglect. The particular discoveries of a particular vendor or shopper, his or her ability to select the most delicious examples, to recognize what subtly distinguishes one piece of fruit or one cheese from another—all of that is copia.

Yes, you should watch out and make sure that you are not swindled. Still, it would be a mistake to believe, as a somewhat racist or snobbish Platonic observer might suggest, that each of these restaurants or market stalls is selling identical products, but tries to distract the consumer from recognizing this through a variety of rhetorical devices. Practices of copia locate the different within the same, locate it in a careful examination of what the object is, how it comes to appear as such-and-such in a particular historical situation. They find "uniqueness" in a place that is not Platonic essence: it resides in the moment of encounter, exchange, and performance, where, especially in the cultivated chaos of the market, in a bustling crowd of people and objects, the status and being of subject and object are literally up for grabs.

Folk cultures in the industrial era have understood that industrial products are not merely "objects" attaining form and power through being fetishized commodities; they are samples of infinity, of infinite variety, which is a source of spiritual insight and enjoyment. For example, consider the phrases used on early Jamaican DJ-ing records, such as *Version Like Rain* or *Rhythm Shower,* which express the idea of an infinite number of copies of a song or a rhythm, something that became actually possible in Jamaica in the early 1970s, when local record producers such as Lee Perry gained access to multi-track recording technology and the techniques of sound manipulation and distortion that we now know as "dub."[32] These technologies allowed them to make endless copies of a tune, and transform them into novel and startling variations, using multi-tracking as a way of making a sound collage. The phrase "version like rain" establishes this process of infinite multiplication as part of a second nature, of technology mimicking the excess and plenitude of nature. But it also establishes a claim on this technology, appropriating it from Babylonian industrial culture, resituating it within Rastafari cosmology and theology, so that this proliferation of copies becomes a life-sustaining and therefore spiritual force. Saying "version like rain" means staking a claim to the right to make, consume, and embody the abundance of copia.

On and On 'til the Break of Dawn

Hip-hop's founding father DJ Kool Herc, a Jamaican expat living in New York City in the 1970s, links the work of Jamaican DJs and producers to hip-hop's emergence; and once again, we find that the notion of infinite abundance, and of mathematical concepts and practices related to it, abound in hip-hop. Houston Baker has written about "massive archiving" as one of the sources of hip-hop's power—the archiving of the "recently outmoded," of records no longer played on the radio or in discotheques.[33] In *Yes Yes Y'All,* early participants in the culture repeatedly speak of the obscurity of the records that Kool Herc plays, and of Afrika Bambaataa's massive record collection. Hip-hop emerges out of this vast garbage heap of postconsumer debris. Hip-hop is also defined by the move from the finitude of the individual recording, played in its entirety, to the infinity of two or more copies of a record, mixed together, edited, and combined in numerous ways. This is equally the case with MCs, who develop their art as a never-ending ocean of wordplay: in the words of the Sugarhill Gang, whose "Rappers Delight" goes on for fifteen minutes: "Well it's on on an' on an' on on an' on / the beat don't stop until the break of dawn." Or how about, "I know a man named Hank / he has more rhymes than a serious bank"? Perhaps you're thinking, Yeah but they're exaggerating. It's true that the boasting and exaggeration that MCs do could fit in well with the outrageous talk characteristic of Bakhtin's marketplace, or, for that matter, with the tradition of signifyin' that Henry Louis Gates Jr. talks about in *The Signifying Monkey.* But in hip-hop culture, these boasts and exaggerations are transformed by being linked to a set of technologies of amplification (the mike), fragmentation and recombination (the DJ), and distribution (the radio, the club, the recording) that can deliver the goods and make actual the possibility of an engagement with the vastness of the infinite.[34]

The first time I taught hip-hop, I asked students why the five elements of hip-hop were called the five elements. There were various replies, but one student told me that in medieval times, there were four elements, earth, air, water, and fire, and people were able to understand the whole universe as being compounded of these elements. Hip-hop, she said, has five elements, and these five elements (MC-ing, DJ-ing, breakdancing, graffiti, and b-boying/fashion) also provide a framework for viewing the universe. I am often frustrated by the way that hip-hop is written about, how narrow the range of reference is—but this student got the scale right. Hip-hop is an extraordinarily vital example of how to make a culture from copying—how to respond to the industrial world with its particular discourses of copying, along with its vast colonial legacies of enslavement and mimetic appropriations of bodies, cultures, and environments, and how to call upon a counterphilosophy of copia (with roots in West African culture, with roots in Bakhtinian folk culture) and make it work.

Hip-hop was born out of gang culture. Which is to say, out of knowledge of the ways in which humans get together and identify as groups. Gangs are an example of mimetic behavior. So is being a student, a professor, a president, a husband or wife, and so on. Within the mainstream of Western culture, gang culture is a new kind of mimetic behavior—a conscious development of a set of shared codes of behavior and being. In his history of hip-hop, Jeff Chang lists the following names of street gangs: Javelins, Reapers, Savage Skulls, Black Spades, Seven Immortals, Mongols, Roman Kings, Saints.[35] The names function as appropriated masks of power torn from global mass culture, vectors of mimetic transformation.

One could track the vast history of "groups with names," from corporations to outlaw groups to nation-states, religious brotherhoods, families, and clans. Indeed, this is more or less what conventional history does. Such a history would be another kind of history of the mimetic faculty—a history less of objects or events than of the building of collectivities, or "publics" in Michael Warner's coinage, with their structures of filiation, identification, and likeness.[36] Hip-hop's predominance across the planet today as an exemplary counterpublic indicates that new kinds of affiliation are emerging. Hip-hop was one of the first places where such affiliations appeared, and also one of the strongest in terms of how many people were drawn toward this affiliation. People from all over the planet now identify with hip-hop, and that identification may be stronger than any other identification in their lives. Although all publics and counterpublics are mimetically constituted, what is striking about hip-hop as a set of cultural practices is how consciously it articulates and negotiates mimetic processes—including those that lead to violence and communal breakdown, which Girard has discussed at length.

Vicious Styles

> One cannot ever splice style—one can only splice segments which relate to a conviction about style. And whether one arrives at such a conviction pretaping or posttaping . . . its existence is what matters, not the means by which it is effected.
>
> —Glenn Gould, "The Prospects of Recording," in *Audio Culture: Readings in Modern Music*, ed. Christoph Cox and Daniel Warner (2004)

One of my students decided to do a presentation for my copying course by inviting a group of breakdancers he knows at York to come to class. In the end, only two showed up, a very tall African-looking guy called Troublez and a stocky Asian guy called Vital. My student began by asking them questions, and I followed. Troublez emphasized the difference between breaking and b-boying: breaking was a formal set of moves and steps (with the implication that breaking was the stereotype or media cliché found in movies and the like . . . the signifier!), while b-boying involved freedom in improvisation. We talked about originality and the question of "styles," which comes up again and again in the hip-hop documentary *Style Wars* (1983), where writers and dancers talk of "vicious styles," "wild styles," and the like.

The dancers differentiated between "power" and "style" in breaking. "Style" refers to a combination of generic breaking moves, like toprock and downrock, within which one brings elements of individual expression and improvisation. "Power" moves are more athletic set pieces. In a battle, power is whatever fancy, stunning move you have up your sleeve, whereas style has more to do with humor, your ability to play with or play your opponents, trick them, parody them, working within the established idiom of breaking moves. I think the fact that these two terms are contrasted is important: "power" belongs to a certain kind of brute force, mastery, which everyone is aware of; it is often quantitative, in the sense that a triple somersault is logically and logistically "better" than a double somersault. "Style" has to do more with the skills of those who lack power—with their power to play, appropriate, trick, copy, to call upon affective and qualitative abilities in order to "move the crowd."

Vital said that, in a sense, there were absolutely no original or individual moves; everything has already been done, many thousands of years ago. "Style" was a matter of the way in which an individual did something, made it his or her own. He pointed out that we were all wearing similar clothes, yet we all brought some individuality to how we wore them, even though they were copies. Style is a way of copying, a way of imitating, and it is this way which can be said to be original. Innovation in hip-hop takes place within a relatively tight grid—you can't just go off and do a bunch of modern-dance moves and expect people to show you respect. Troublez pointed out that when two dancers battle, they are responding to and imitating each other's moves. They might parody, repeat, play with elements of each other's steps—although what actually happened when Troublez and Vital danced was that rather than battling it out, they ended up standing firmly on their own ground and style, so that there was little possibility of even comparing them.

Vital was very circular, incorporating a number of moves from martial arts, yoga, maybe even modern dance. He touched his foot to his head, stood on the side of his foot. Troublez, tall and lanky, flew around, all angles, long legs and arms shooting straight out, before going into an amazing head-spin, pirouetting on his head for forty-five seconds. I'd never seen anyone do moves like this before; but I would say that Troublez was more about power, Vital more about style. Still, it was a clear draw, with no obvious winner. Troublez concluded by noting that what made b-boying different from Tae Kwan Do was that in Tae Kwan Do there is a form and that's it—you have to follow it. Whereas hip-hop had "freedom of expression," and although it was capable of incorporating elements from just about anything, it also allowed one to innovate through the ways one used the elements and combinations, and introduced "new styles." Hip-hop, then, is exemplary in providing a set of practices or frameworks through which a variety of more traditional folk cultures and forms—martial arts, dances, warrior rituals, poetry, filiations—can meet and take on new form without losing the particular history from which any individual or local group might engage with it.

Of course, "style," including hip-hop style, has long been integrated into the capitalist marketplace—and there could be no capitalist market at all without very particular economic organizations and appropriations of copia's abundance.[37] But to see style, and for that matter "copying," as mere epiphenomena of capitalist production is to invert things, and to radically underestimate the power of these forces. The power of hip-hop, and the five elements, which are five "styles" of being in the world, constitute five types of magic, if you like—five ways of transforming things, and therefore five ways of changing what gets called a "person" and what gets called a "world." I would like to think, though I can't prove it, that folk cultures have always had this power, have always discovered it for themselves, insofar as folk cultures are always cultures to whom nothing belongs, from whom everything is taken.

This power of style, always on the move, is a power of "copying," and of copia as abundance. It is predicated on the knowledge that there are an infinite number of ways of presenting, perceiving, disassembling, and reassembling objects and selves, and that this activity is a collective one. Vicious styles are a rhetoric, a rhetoric of performance and performativity, a rhetoric for the production and presentation of both subjects and objects, and the construction of shared worlds. The users of such styles know how to work with things that don't belong to them, and they realize that nothing belongs to anyone—that everything has already been done many times in the past, but that every moment is in some sense unique, and to be newly fabricated.

Copia and the Sovereignty of Folk Cultures

Today, words like "subculture" and "subaltern" describe the impasse of folk cultures in the age of global capital. As is well known, "subcultures"—including "indie," "alternative," and "hip-hop"—have allowed themselves to be appropriated into mainstream consumer culture to the point of almost total cooptation. The great graffiti artist Rammellzee lamented the fact that in the early 1980s, graffiti crews, who established their sovereignty over the city of New York by writing burners across entire subway trains, traded this sovereignty for "subculture" and a chance to participate in the international art market. But the sovereignty of graffiti crews or folk cultures is almost an oxymoron—since they are mostly defined by their marginality.

There have been a number of attempts to articulate the collective political potential of subcultures. Such radicalization necessarily involves reasserting the connection of contemporary subcultures with the broader flow of folk movements and forms—for example, Paul Gilroy's theorization of the Black Atlantic as a counterculture of modernity, and the recent elaborations of Afrofuturism as the digital global dissemination of Black Atlantic practices.[38] One could also extend Greil Marcus' reading of the politics of Bob Dylan's turn toward "the old, weird America" and folk forms, beyond the American countercultures of the 1960s, in the direction of a more global folk turn that would include Brazilian Tropicalia, and various European and Japanese freak and anarchist cultures and communities.[39] Punk, whether discovered today in Shanghai or other Asian metropolises, or Islamic taqwacore as it is heard in various places around the world, also reveals itself as a potent, global, countercultural form.

Yet a certain aesthetics of failure, indifference, idealism, or perversity in relation to the official marketplace is one of the characteristics of the participants in such cultures—"the curse of the Fall," as Mark E. Smith described it—and the rest of us, too. The reclaiming of copia is a part of the radicalization of any subculture, and it manifests itself historically in legal struggles concerning alternative economic modalities, in an aura of inauthenticity, and in a politics of refusal. There's nothing too pure about any of this: it's not about benevolence—participants in folk cultures steal other people's styles and incorporate them, and are themselves stolen and incorporated. They are suspicious of art, and often see themselves as workers for hire, even when this work requires a high degree of aesthetic or technical sophistication. And they're often tangled up with gangs, mafias, gray markets, which are pretty ruthless about the bottom lines of power and money. This might indeed be "the community of those who have nothing in common"—except copia. But maybe that is enough, since, as Ernesto Laclau observes in his recent work on populism, attempts to describe a concept of the popular and the people are usually warded off with "accusations of marginality, transitoriness, pure rhetoric, vagueness, manipulation."[40] In other words, the groupings of "the people" are already considered derisively or dangerously mimetic—and what is needed is to affirm the autonomous and skilled (in Laclau's terminology, "reasoned") way in which folk cultures constitute themselves through mimetic processes as a way of coming to power and collective joy.

One of the most interesting recent attempts to work through these issues has been undertaken by a group of musicians from Seattle who go by the name of the Sun City Girls, and who now operate a record label—Sublime Frequencies—specializing in global contemporary folk musics.[41] The group is controversial precisely because of their refusal to present global folk musics as isolated ethnomusicological specimens; they prefer to imagine them as elements of a fragmentary, anarchic collective which stages a "carnival folklore resurrection," to use the title of one of their records. Their presentations of these musics are often unattributed montages made from recordings of local radio broadcasts during their global travels; at other times, they collect recordings purchased from street cassette vendors in various cities of the global South. Conversely, they have also recorded cover versions of various folk musics which they have then anonymously inserted into the stacks of the same street cassette stalls during their travels. This is a crazy, utopian punk-rock-inflected project—but it precisely expresses the way in which punk rock was and is able to articulate an ethos and practice of copying that envisions different economic forms, different forms of community and justice, beyond those negotiated in liberal intellectual-property regimes.

Predictably, one of the main consequences of this activity, which continually evokes the excess of copia (titles such as "Box of Chameleons," "Folk Songs of the Rich and Evil," an endless flow of limited-edition cassettes, LPs, and CD-Rs, all of quasi-official status), has been that Sublime Frequencies / Sun City Girls are regularly accused of copyright infringement. Without denying the right of any person or collective to equal access to existing structures of intellectual property such as copyright law, is it not the case that any attempt today to "constellate" a people, in Laclau's sense of the word, will find itself blocked by IP regimes that refuse such nascent collectives the right to present and articulate themselves outside existing property regimes? Of course, they can be "inserted into hegemony" (as Spivak puts it) and become "owners" too. But localized folk collectives, such as street cassette vendors in Southeast Asia, or users of fileshare programs around the world, already inhabit a vastly expanded public domain which is enacted in the systems of exchange that they are involved with. What is needed today is an expansion of the space and opportunities conducive to the mutual appropriations of folk cultures, for copia involves a movement of forms and energies that is antithetical to that of private property. It is open, unobstructed, and—from the point of view of form—inherently multiple, excessive, and abundant.

3/Copying as Transformation

Chuang Tzu and the Butterfly

The Chinese philosopher Chuang Tzu tells the following story: "Once Chuang Tzu dreamt he was a butterfly, a butterfly flitting and fluttering around, happy with himself and doing as he pleased. He didn't know he was Chuang Tzu. Suddenly he woke up and there he was, solid and unmistakable Chuang Tzu. But he didn't know if he was Chuang Tzu who had dreamt he was a butterfly, or a butterfly dreaming he was Chuang Tzu. Between Chuang Tzu and a butterfly there must be some distinction! This is called the Transformation of Things."[1]

What does this story really mean? At first sight, it seems that we are being asked to choose between the butterfly in the dream and the embodied Chuang Tzu, between the real world and the dream world, which suggests that both are equally valid. Chuang Tzu is saying that both are equally *invalid*—and that both "butterfly" and "Chuang Tzu," "dream world" and "real world," are ultimately only conventional terms. Transformation is constantly taking place as we shift in time and place. Everything is continually moving. The labels that we place on these movements are conventional, and produce the appearance of solid things such as a butterfly, a dream, a Chuang Tzu, and a real world. Underneath these names, the slow chaotic work of becoming and transformation is constantly taking place. Yet at the same time, the recognition in the story takes place only because of the words, and Chuang Tzu does at some level identify with his name, as well as with the butterfly he feels he is when he dreams. "Transformation" happens only because there is something named a "butterfly" that can be compared and contrasted with something known as "Chuang Tzu." Without these powerful acts of naming, nothing called "transformation" could be identified, and no "distinction" either.

Chuang Tzu is suspicious of words, labels, ideas, because their representational status cannot track the constant flux of the cosmos. Nevertheless, there are plateaus of being in his writing, and a sense of times, seasons, epochs in which things manifest in particular ways. He is too intimate with nature not to recognize the cyclical, relational value of words like "birth," "youth," "old age," "death," or "spring," "summer," "fall," "winter." Nature is not just a total flux; it also involves homeostatically balanced, repeating displays that endure for a certain period of time, and these states and entities are worthy of being named, even if they conceal enormous undertows of flux and change. This is also the case in the classical Chinese text known as the *I Ching,* or Book of Changes, which constructs a cosmology by naming and marking off sixty-four of these plateaus and the cycles and paths that are possible between them. Wisdom, at one level, consists in acting in accordance with these states, which means allowing oneself to be produced, as entity, as outward appearance, in accordance with a particular configuration of natural forces, as they arise in a particular time and place. In his interpretation of the

Chuang Tzu text, Jacques Lacan astutely points out that while the waking dreamer can wonder whether or not he is a butterfly, the butterfly in the dream does not wonder whether or not he is a man who is dreaming he is a butterfly. Slavoj Žižek, in his gloss on Lacan's interpretation, warns us against reading this story as signifying a postmodern essenceless floating subject, since not all identities and identifications are the same. For Lacan, the butterfly in the dream is "a butterfly for nobody," whereas "it is when he is awake that he is Chuang-Tzu for others, and is caught in their butterfly net." But neither of these milieus has an absolute essence either; they are contingent, interrelated, and impermanent.[2]

In previous chapters, we established that copying, rather than being the production of a distorted, inferior version of an original, emerges from emptiness and from the impermanence, dependent origination, or lack of essence of all things. This impermanence, which in Chapter 2 we described as a network of infinite, connected, essenceless signs with the name of Copia, is always already in a state of transformation; and upon examination, no conceptual apparatus of any kind can describe it—it is utterly beyond concepts, although words like "suchness" and "sameness" may point to it provisionally.

Copying is a particular kind of transformation, yet it is very difficult to think of it in this way. Consider the following: if the copy of the book you are reading is a "hard copy," its matter was once a tree. If you are reading an electronic copy, then the matter and forces that make up the pixels of the screen have gone through even more radical processes of transport and transformation. Yet despite our increasingly sophisticated understanding of ecology and the way that particular forms arise from and in dependence on a network of others, the illusion that things are self-produced and separate from one another remains powerful. We are fascinated by our own ability to transform things, including ourselves, through imitation, yet we forget that such processes of mimetic transformation are continually and universally operative, rather than occurring only when we "make a copy" or consciously imitate something. What would it mean to live in awareness of this knowledge—that everything is in the situation of Chuang Tzu and the butterfly? And conversely: Why do we remain consistently unaware of it?[3]

A Human Chameleon

Woody Allen has been concerned with copying throughout his career, beginning with his early movie *What's Up Tiger Lily?*—in which he appropriated a 1960s Japanese gangster movie and overdubbed a new soundtrack to the film, all dialogue in English. *Zelig* (1983) is a faux documentary concerning one Leonard Zelig, a Jewish New Yorker who comes to fame in the 1920s when it is discovered that he compulsively imitates whatever social milieu he finds himself in, transforming himself both physiologically and mentally into a reflection of the people who surround him—people of various ethnicities and professions. Like Franz Kafka's "Report to an Academy," in which a former ape recounts the process by which he transformed himself into a human being, *Zelig* is in part a sly look at the paradoxes of Jewish identity in twentieth-century Europe and America, at the problems of assimilation and persisting stereotypes of Jews as duplicitous imitators, hiding secret agendas behind a façade of integration into the societies they live in.

Zelig is a compulsive imitator, and at times it seems that there is nothing stronger holding his psyche together than his mimetic abilities. The key scene in *Zelig* in this regard is the one where Eudora Fletcher, the brilliant young psychoanalyst, attempts to treat Zelig through hypnosis and therapy, at her retreat in the Hamptons, while her brother films their sessions. Zelig proves highly resistant to this treatment, transforming himself into a doctor, reflecting her own "outward appearance" back to her through his own distorted copying—the distortion marked through parodic humor. Fletcher takes a break and goes to the city, where she watches a theater performance and kicks back at a nightclub full of dancing girls. Returning to the Hamptons, she starts her next session by confessing to Zelig that she is not a doctor, imitating his own imitation of her. He visibly crumples as the solid basis for his own performance disappears. He is rehabilitated and individuated through his love for Fletcher; but under stress, the compulsion to imitate breaks through again.

One simple way to put it is that a copy is a repetition. Zelig repeats, with his outward appearance, the outward appearance of the environment that he is in. Zelig's differentness is revealed only when his own transformation is repeated back to him. In *Difference and Repetition,* Gilles Deleuze argues that we can understand repetition only "once we realize that variation is not added to repetition in order to hide it, but is rather its condition or constitutive element, the interiority of repetition par excellence."[4] Repetition contains difference within it, just as copia necessarily involves variation in the constitution of what we call "copies." And, to reintroduce once again the despised but unavoidable rhetoric of holism (a rhetoric irreducible to any *conceptual* holism), difference is produced through nonconceptual sameness or suchness or nondifference.

The central conceit in *Zelig,* however, is that Zelig's compulsive imitation is a reflection of an almost universal "compulsion to become similar," which Europe and America manifested in the 1930s and are still experiencing. Once he is tracked down and arrested after leaving his workplace, Zelig becomes an immediate object of scientific scrutiny, appropriated by medical specialists of every kind as an example of their pet theories. Both the examinations and therapies that Zelig undergoes transform him in radical ways: one therapy leaves him with his legs twisted around 180 degrees; another has him walking the walls like the bug in Kafka's tale "The Metamorphosis." He quickly becomes a media celebrity too, his every movement reported in the newspapers. Songs and dances are named after him; watches, games, and puppets bear his name or image.

Zelig's similarity, it seems, can be transferred to absolutely anything. It appears highly contagious: his own compulsion to copy is compulsively copied by those around him. The effect is dizzying. What emerges is the picture of a society in which the will to copy is all that holds things together. Zelig becomes the stereotype for this generalized urge—but the identities of doctors, media figures, workers, armies, politicians, and others appear to be no less a product of the desire to imitate. Zelig is, to use psychoanalyst Bruno Bettelheim's formulation in the movie, "the ultimate conformist." But the masses, whether they imitate Zelig, the American news media, Hitler, or the Communists, are no less conformists. And isn't Zelig's period of happy individuation, where he acts like a normal person or a celebrity, in itself conformist—and mimetic? Yet "conformism" is only one way of thinking about this compulsion to imitate—the loving pastiches of popular songs, fads, and crazes in *Zelig* celebrate the joy of transformation. Zelig becomes a figure of fascination because people recognize in him their own desire to transform. He becomes an occasion, even an excuse, for people to transform themselves.

Universal Imitation?

Is our society held together by nothing stronger than the compulsion to imitate, and to transform through imitation? The philosopher who has made the broadest claims in this regard is the French founder of sociology, Gabriel Tarde, who, in *The Laws of Imitation* (1890), sets out the three principles of universal imitation: a vibratory principle operating at the physical level; a reproductive principle working at the biological level; and imitation working at the social level. All phenomena result from repetition at these various levels, and from the interpenetration of repetitions, which give rise to variation and innovation. It is a beautiful theory, but Tarde stumbles when he inquires as to the nature of the primary substance of the universe—that which would by definition be nonrepetitive, since it is original.[5] The great Atomist philosopher Lucretius solves this problem by speaking of a *clinamen*, an inherent swerve in the basic constituents of the universe as they move in the ether, a swerve that results in the variety of the universe as we know it. Perhaps surprisingly, there have been a variety of Buddhist atomisms too, notably that of the seventh-century Indian Buddhist philosopher Dharmakirti, who proposed that there were infinitesimally small partless particles whose interaction with mind resulted in the phenomenal world. From a strict Madhyamaka perspective, there can be no such original or primary matter from which the universe is composed, and the label "atom" can be only a relative, relational concept that is itself subject to impermanence and transformation. Even within this Madhyamaka perspective, however, Tibetan interpreters of the great Indian Buddhist philosophers produced quite radically divergent ways of understanding phenomenality.[6]

Where does the word "copy" stand in relation to Tarde's imitation through repetition? The word "copia" is connected via its Latin root to the word "copula," meaning a tie, a band, a fetter.[7] And therefore to the word "copulate." At one level, we all know that sexual reproduction involves copying, but there is an obscenity in this recognition before which the mind recoils. The obscenity relates to the element of repetition in both copying and reproduction: the repetition in the act of making a copy, whether by making love or by other means, in which pleasure is connected to a disappearance of self; and the repetition in the copy itself, which, in the form of the individual baby, is both different from and the same as all other babies, human or otherwise.

Sexual reproduction is of course at the core of the post-Darwinian modern biological understanding of life on earth. Molecular biologist François Jacob argues that there is no such thing as a living system without reproduction, without the "desperate eagerness" of an organism to replicate itself. In the opening sentence of his book *The Logic of Life*, he says: "Few phenomena in the living world are so immediately evident as the begetting of like by like." Yet Jacob himself, despite noting the existence of plants that reproduce themselves "without changing" for millions of years, does not use the word "copy" except to say that "the reproduction of a living being . . . does not simply involve making a copy of the parent at the time of procreation."[8] Organisms reproduce themselves, and Darwinian natural selection necessarily involves the production of variants that are said to possess individuality, difference, rather than being identical copies. Only in the margins, peripheries, ghettos of life, where the status of "living" is in question, do we speak of copies of organisms—with reference to viruses, bacteria, fungi, microscopic organisms, vermin. Or in the discourse of eugenics, where humans and others are relegated to the status of the nonliving.

In the strict sense of the term, though, sexual reproduction through the transmission of genetically coded traits is copying, and the conventional Darwinian argument to the contrary is an ideological one. A new being is produced by this act of copying, whether through the blending of the genomes of mother or father, or, in the case of the asexual reproduction of certain bacteria, through mistakes in the transcription of a genetic code that should otherwise be perfectly repeated. The copy is not an identical copy—but then, as we will see, it never is. This creative act of copying is never spoken about as such, because we are uncomfortable with the Platonic associations and the social/political/biological discourse that supports it. Who, after all, would call their offspring inauthentic or an imitation?

Copying as Reproduction Envy

The word "copying" evokes images of gadgets, technologies of mechanical reproduction, or the masterly hand of the artist who is particularly skilled at producing reproductions. It is a stereotypically masculine activity. Is copying, defined thus, a specifically male attempt to imitate, appropriate, fix, and control the knowledge of becoming and transformation that is a part of women's experience of their bodies, through the menstrual cycle, pregnancy, giving birth, and nurturing a child?

If we took pregnancy and childbirth as our model for mimesis, we would have to frame things quite differently. In contrast to the ejaculatory male discourse of copying, summarized nicely by Freud in his notion of the pleasure principle, we would need to explore a nonappearance of the copy that is still felt very viscerally—the kicks, movements of the baby in the womb, but also the hormonal, chemical, nutritional, corporeal, ultimately ontological shifts that the expectant mother experiences. Concealing and unconcealing, which Heidegger points to as the rhythms of Being, are aspects of a mimetic process. The unborn baby is relatively undifferentiated from the mother's body; it emerges out of a sameness, yet evidence of this emergence takes the form of signs of differentiation which the mother is aware of. The mimetic process continues, after the unconcealing of birth, with breastfeeding and with the ongoing need for intimacy with the mother's body. All of this is well known. But what if such a mimesis was actually the primary state, a stateless state of becoming—a slow/fast rolling wave from conception through birth, "being," death, and onward that never ends, and where "outward appearance," in the sense of a particular fixed identification, was, finally, deceptive?

The film *Being John Malkovich,* scripted by Charlie Kaufman, explores the tension between such masculine and feminine models of mimesis. Craig is a down-at-heels puppetmaster who puts on puppet shows in which his everyday life fantasies are sublimated. His partner, Lotte, looks after a menagerie of animals and works in a pet shop. She wants to have a child; he doesn't. At his new workplace, where he lusts after his co-worker Maxine, Craig discovers a portal into the brain of actor John Malkovich. By entering this portal, located behind a filing cabinet, anyone can share Malkovich's embodied sensory experience (no one seems too interested in his consciousness or what he is thinking!). Rather than "being John Malkovich," those who enter the portal are, we might say, allowed to temporarily inhabit the outward appearance of John Malkovich from inside (the camera looks out through his eyes during these sequences). At first this takes the form of passive observation of Malkovich as he goes about his daily routines. But very quickly, it becomes a more active appropriation of Malkovich's will and consciousness, so that the inhabitants are able to use the outward appearance of Malkovich for their own purposes. As an actor, Malkovich already is assumed to have a certain mastery of mimesis, and as a celebrity he is also the object of intense but banal mimetic identifications of various kinds.

While Craig is fascinated with the "metaphysical can of worms" that the portal opens, Lotte has a profound experience of embodiment when she enters Malkovich's body and decides that she wants to undergo gender reassignment. Meanwhile Maxine, who displays no interest in entering Malkovich, makes a date with him. Both Craig and Lotte are in love with Maxine, and Maxine falls in love with Lotte while inhabiting Malkovich. Lotte and Maxine conduct a torrid love affair, through the body of Malkovich, until Craig discovers what is happening and locks Lotte up in an ape cage. Craig then forcefully takes over Malkovich's body and, using his puppeteering skills, seduces Maxine. They become a couple, and Maxine becomes pregnant by "Malkovich." Things unravel, though: Craig loses control of Malkovich, and Maxine and Lotte get together, in their "own" bodies, to raise Maxine's child.

Craig has an instrumental view of copying: puppeteering is his career, and he makes copies of things in part so that he can manipulate them to his advantage. A copy remains a thing for him, even when he invests very intensely in it. With this intense investment comes a technical mastery that has a certain beauty to it. When things do not go his way, Craig resorts to violence as a way of forcing things to happen in the way he wishes. Although he transforms external objects, he himself is not transformed.

Lotte, on the other hand, is overeager to transform, and makes empathic investments both in the animals she looks after and in the body of Malkovich, which she immediately (and wrongly?) experiences as her own. As with Craig, transformation for her is a mode of escape, but unlike Craig, she also sees it as a mode of becoming: she becomes herself through "being John Malkovich," and at the end of the movie she and Maxine get together, apparently in acceptance of who they actually are.

Maxine is the most complicated character in the movie. She has no interest in transforming herself through becoming Malkovich, yet she becomes pregnant with Malkovich's baby, which she regards as Lotte's because Lotte was inhabiting Malkovich's body at the moment of conception. The decisive transformation in the movie, then, is not Craig's technical mastery of transformation or Malkovich's biological right to be named father, but the change caused by the love that Lotte and Maxine feel for each other, which is capable of overcoming even a biological barrier to their mimetic transformation.

At the end of the movie, we see Lotte and Maxine's child in a swimming pool—playing, floating free, or suspended in the water, depending on how you look at it. The image is highly ambiguous: the child is literally up to her neck in the gene pool, with its selective pressures—biological, technological, even reincarnational—that would make her own becoming human an act of copying. Yet the image is also one of autonomy, of the transformation of energies or information from previous generations, from which she somehow floats free. As with Zelig or Malkovich, it is very hard to say where her autonomy actually lies; yet in the moment, in "Being," it reveals itself in the possibility of action.

Lovemaking as Transformative Mimesis

In his analysis of mimetic desire, René Girard focuses on the importance of the triangle: the model who desires; the object who is desired; the rival who is contagiously infected by this desire and imitates it, leading to an escalation of rivalry in which the object is forgotten, replaced by the mutual fascination between model and rival. This fascination leads to a cycling of mimetic energies between the two as they exchange and repeat each other's gestures, often leading to violence, in a desperate attempt at producing a decisive differentiation.

But what happens when there are two people who are the object of each other's desire? Without foreclosing the possibility of all the others who are involved in desire (quite literally in the case of *Being John Malkovich*), is it possible that lovemaking reveals an undisplaced mimetic power, with the attendant possibility of biological consequences in terms of reproduction? Yet reproduction is not a necessary consequence of making love—a fact that indicates the plasticity of mimetic energies, the ease with which they are diverted from the biological to the social and/or physical levels and back again, aiming at more immediate states of undifferentiation, of sameness. Not the sameness that Freud saw as the basis of the pleasure principle, that deathlike state of satisfied neutrality following the male orgasm, when one sleeps or stares thoughtfully at the ceiling. Instead, it is the sameness of nonduality, the undifferentiated that is beyond concepts, dynamic rather than homogeneous or neutral. Motion is not necessary for this nonduality, which is in fact the only thing that is "always already" there; but in lovemaking, we point this presence out to each other and then we realize it, temporarily, in orgasmic jouissance.

How exactly are we to understand mimesis in relation to lovemaking? What interests me is the trajectory that leads up to "sexual reproduction," for the whole of this trajectory must be considered a mimetic phenomenon. This would include the repetitious movements of the body, of the breath, of vocal sound; touch, and all the complex feedback loops of tactility that happen between two people in motion; the visual perception of the object of desire; the gestalt or environment of the erotic encounter, which tends to be overdetermined either in its familiarity (the bedroom) or novelty (the kitchen floor, the public bathroom, etc.).

We then have to consider something extraordinary: that mimesis is not just a matter of a particular situation in which the outward appearance of something is changed so it becomes similar to something else, or an invisible contagion capable of causing such a change—but that everything in this world, insofar as it "is," "is" because of the transformations that mimesis makes possible. Whether such change be the slow transformation of a particular body and mind from a baby to a child to an adult to an old person to a dead person, in which it is the "same" person although this sameness is possible only due to a continuous cycling of living cells fueled by food and air and everything else; or whether it be the more rapid transformation by which, through sexual intercourse, two beings produce a third that gestates in the mother's womb for a certain period of time before coming forth as a separate, "different" entity; or whether it be the still more rapid transformation of a piece of stone or clay into a statue of that baby. Everything is shifting, impermanent—"*mimetism* itself," in the words of Philippe Lacoue-Labarthe, "that pure and disquieting *plasticity* which potentially authorizes the varying appropriation of all characters and all functions . . . but without any other property than an infinite malleability: *instability* 'itself.'"[9] "Copying," then (and this is something I will take up in the next chapter), is a sign of or symptom of or scapegoat for this primordial plasticity of name and form, to which Lacoue-Labarthe gives the name "mimesis," fully aware that in naming it he is representing and thus freeze-framing the instability.

Lacoue-Labarthe goes on to say that "what is threatening in mimesis is feminization, instability—hysteria" (129). He cites Luce Irigaray briefly, but does not pursue the topic. Yet Irigaray speaks at length of what she calls "maternal sameness," which she defines in terms very close to those of nonduality, as a similarity beyond concepts. "This sameness is not abyss; it neither devours nor engulfs. It is an availability so available that for one who lives for utility, for mastery, the cash nexus, debt, this assumption of availability—which precedes any position that can be discerned—arouses anxiety and hence efforts to name and designate causes. This sameness is the maternal-feminine which has been assimilated before any perception of difference."[10]

Being John Malkovich finally affirms this "maternal sameness" against all the male gimmicks, ruses, and stratagems of "copying" which seek to both suppress and supplant it.[11] Yet those copies also emerge out of it. Irigaray's "availability" also suggests the "infinite malleability" of mimesis—an availability that underlies all transformations and every thing that comes into being.

There is a profound relationship between mimesis and the feminine which is obscured by conventional discourses of copying. We speak of "matter," of "materialism," of cosmos as matrix, of emptiness as "Mother of the Buddhas"—all of these terms being derived from the maternal. Even the word "bag," as in "Louis Vuitton bag," comes from the Sanskrit word *bhaga,* one of whose meanings is "womb."[12] In absolute terms, nonduality is beyond all concepts, including those of gender. There are, however, sites, situations, beings, events with a privileged relation to the nondual, since they model or instantiate it, including the maternal-feminine, whether in a mother looking after a baby, or the drag queen House rulers of *Paris Is*

Burning, with their crews of wild youths, where care for the other manifests through a perception of intimacy, of nonseparateness and the recognition of the dependent origination of all beings.[13]

From Imitation to Transformation

To repeat: in every case that one can think of, copying involves repetition. Repetition—a copy repeats, is a repeat of something. But in this act of repetition, as Tarde and others have suggested, something else happens. Difference manifests itself in repetition and marks a transformation that happens within repetition. Thus the man, the butterfly, the waking state, the dream: some continuity between them has to be assumed, even when the discontinuity is so apparent. Those that slide down the manhole for their fifteen minutes inside John Malkovich are looking for a self-transformation that comes when their outward appearance is that of John Malkovich, but they also trust that it will be their own consciousness that repeats itself, that continues, inside John Malkovich. Total transformation would mean complete self-erasure in John Malkovich, and then there could be no enjoyment of the event. On the other hand, when Malkovich enters his own brain, he is capable of seeing the world only as an endless collection of repetitions of Malkovich, both as name and form.

Elias Canetti argues in *Crowds and Power* (1960) that imitation is only the first stage on the way to total transformation.[14] He is fascinated by surfaces and depths, and notes that an imitation may involve merely the repeating of certain superficial characteristics, while in the depths everything else remains the same. At the other extreme, transformation involves the changing and becoming of the totality of all that constitutes a particular entity or form—in human terms, the mind and the body. Possession is interesting from this point of view, since it entails a transformation of consciousness, a transformation in the depths, that also manifests at the surface level, as we see in a movie like *Divine Horsemen: The Living Gods of Haiti* (1985), by Maya Deren, Cherel Ito, and Teiji Ito. The body moves differently; the shape of the face changes, and the voice too. Yet for all that, it is the same body, and the transformation remains temporary, incomplete. We might also speak of powerful drug experiences in which consciousness is profoundly shaken or transformed, but at the same time this consciousness in some sense persists. The transformation is temporary and at some level continues to manifest signs of the original. So total transformation is very rare, in Canetti's sense. I am struggling to come up with even a single example. At the point where a being is totally transformed, can we even speak of transformation? Doesn't there have to be some trace of the original? A baby grows up into an adult and then dies. Is death, then, this event of total transformation? Perhaps. But even then, according to the Buddhists, the transformation is only partial. Some trace persists; necessarily so.

Canetti also speaks of a stage between imitation and transformation which he calls "simulation," or its counterpart, "dissimulation." By this he means the deliberate imitation of certain qualities to disguise one's true intentions. Simulation: one changes one's surface, while inside one remains the same. Dissimulation: one maintains one's surface, while inside something has changed. Thus, the dictator wears various masks, such as those of civility and politeness, while inside he or she is ruthless, unchanging. Conversely, those that plot against the dictator act as though they are loyal subjects. The dictator would like to unmask them all. He is paranoid. And he may or may not have the power to unmask.

Violence

There is a tremendous energy encoded and encased in a particular limited, defined form with a particular name and identity. If we acknowledge that nothing stays the same in this world, and that the maintenance of a particular form requires certain kinds of violence and appropriation—for example, that living beings need to eat other living beings (including plants) in order to survive—then our insistence on the right to maintain a particular name or form is not unproblematic. According to Mircea Eliade, medieval alchemists "projected on to Matter the initiatory function of suffering" and recognized that their violations of matter amounted to a form of torture whose end result was transformation.[15] Contemporary Vedanta teacher Sri Nisagadatta Maharaj once responded as follows to a questioner demanding to know whether "existence and conflict are inseparable":

> M: You fight others all the time for your survival as a separate body-mind, a particular name and form. To live, you must destroy. From the moment you were conceived, you started a war with your environment—a merciless war of mutual extermination, until death sets you free.
>
> Q: My question remains unanswered. You are merely describing what I know—life and its sorrows. . . . Give me the final answer.
>
> M: The final answer is this: nothing is. All is a momentary appearance in the field of the universal consciousness: continuity as name and form is a mental formation only, easy to dispel.[16]

What is striking in this exchange is the insistence on the ubiquity of violence at the relative level which dissolves into nonduality at the ultimate level.

The most thorough exploration of violence and mimesis has been undertaken by René Girard, for whom acquisitive mimesis, the desire to appropriate something that belongs to another, is a crucial feature of both animal and human life.[17] While animals manage situations of otherwise "undifferentiated violence" through an "instinct" that sets up hierarchies within a group of animals, human beings are distinguished, according to Girard, by their lack of such an instinct. Situations of mimetic rivalry for an object inevitably escalate to a point where violence threatens to spread contagiously through a community until the community is destroyed. Rituals of sacrifice, scapegoating, and other manifestations of a universal victimage mechanism serve as a channel for these otherwise unmanageable mimetic/violent energies. This channeling also unites the community against the victim, whose sacrifice becomes something sacred or transcendental, even as it also results in his or her death. Such a sacrifice or scapegoating also serves to stabilize the identities of all concerned, at least temporarily. Ironically, Girard's distinction between animal and human is itself "mimetic" in his own sense of the word, and serves to make animals the scapegoat of a "human community" that can thus constitute itself in opposition to the rest of the natural world, at the same time rationalizing its own violence against it.

Is violence inextricably linked to the abundance of copia that I set out in the previous chapter? In the Ovidian myth of the horn of plenty, the cornucopia is produced in the heat of battle, in an act of tearing. The long histories of sacrifice, of warfare as a break in the rules governing the distribution of things, of capitalism as the vast-scale appropriation of things and people into the marketplace—all are histories of mimetic violence. And the same preoccupations are evident in hip-hop: the violence of slavery, of appropriation from Africa; racism in America; the destruction of city neighborhoods by Robert Moses; the thrilling violence of fast cars on the Cross-Bronx Expressway, the violence of turf wars and teenage warrior games, battles—all of this was a training in appreciation of montage as an art form, of appropriation as cultural and survival strategy through which flows of abundance are produced, of violence diverted into creative as well as destructive acts. An education in the mimetic faculty, then. Hip-hop has turned out to be a most artful response to the violent abundances of modern industrial society: five elements, five ways of responding to the industrial universe—and copying, in the sense of copia, is crucial to all of them. That being said, hip-hop itself is associated with violence, and the creative aspects of hip-hop have also been channeled back into the violence of gangsta warfare, gladiatorial combat played out on television and on the streets as spectacle and genocide.

More generally, power's discursive repertoire—the architecture of palace, church, and government, the rituals and performative gestures of the state, the statues and other figurative symbols of "greatness," Louis Althusser's ideological state apparatuses—are concretized, given the semblance of permanence, through mimesis. In 1938, Bertrand Russell defined power as "the production of intended effects."[18] Thus, he defined power in terms of mimesis—as the perpetuation of a structure that one imposes through one's will or through a force which must be maintained in order for that structure to stay the same. If, as Michael Taussig writes, we show a lack of reverence for power by defacing the images and structures through which it is expressed, all the ferocious energy encrypted in those forms leaps out—usually in the form of the law, the police, the repressive state apparatuses which are the guarantors and protectors of the mimetic figurations of the state.

While Girard's observations about the prevalence of mimetic phenomena are persuasive, his assertion that violence is at the origin of the human obsession with mimesis significantly undervalues the creative aspect of mimesis. If the need to control violence is at the basis of human institutions that set limits on or structurings of unbounded, contagious mimetic energy, one should inquire into the source of what is called "violence"—something that Girard does not do, beyond saying that it is the consequence of disputes over an object. Violence would then appear as a particular way of labeling, framing, or imitating the transformative mimetic energies of which humans and the world are composed—in other words, a way of appearing to control, to master, the impermanence not just of nature, or human social relations, but the subject's own sense of itself in relation to others. Violence as directed human or organic life action is not just a matter of "survival," in the Darwinian sense, but an aggressive attempt to fix this impermanent flux through imitating it in an apparently controlled way. In particular, it strives to fix a permanent self as existing separately from this flux through the projection of the flux onto an exterior other who is deformed by violence while, optimally, the violent subject experiences his or her own continuity through the agency of an act of violence.[19]

Taboos on Transformation and Copying

Canetti writes about prohibitions on transformation, and connects sexual prohibitions with the transformative potential of sexuality and sexual acts. He branches out from the largely predatory definitions of transformation that he otherwise focuses on, and discusses sexual and shamanistic examples of transformation that suggest an almost universal will to transform. He concludes that "even this brief enumeration of a few instances of prohibition on transformation, which leaves almost everything still to be said, forces one to ask what this prohibition really signifies. Why does man want it? What deep need repeatedly drives him to impose it on himself and others?"[20]

We know from Eliade's book *The Forge and the Crucible* that the violation involved in effecting a transformation of matter was recognized in many traditional societies via a set of rituals, taboos, and rules that governed the process of transformation—from the act of taking matter (say, metal ore) out of the earth, to shaping and transforming it, to using it. Sexual continence, prayer, even human sacrifice were seen as a necessary prelude to the reshaping of matter; and the miner, the smith, the forger, the alchemist were regarded as types of shaman, honored religious figures. In other places, the smith and his family were considered taboo, unclean; and contact with them could bring death, disease, or other misfortunes.[21] With alchemy, the transformation of matter and the consciousness of the smith and alchemist were considered reciprocal or mutually contagious: "As far as the Indian alchemist is concerned, operations on mineral substances were not, and could not be, simple chemical experiments. On the contrary, they involved his karmic situation; in other words, they had decisive spiritual consequences."[22] There was a mimetic link between matter and consciousness, and in their transformations too. From this emerged an ethics—not to say an ecology—of transformations of matter. The advent of a desacralized scientific chemistry in which such consequences and ethics were ignored, and the later integration of this chemistry into industrialized production, enact their own torture on the earth, which can now be plundered for raw materials at will—with further consequences, as we know.

If the taboos on transformation are numerous, then can we speak of taboos on copying? The word "taboo" may seem too strong to describe the prohibitions that exist around copying, yet the word illuminates the irrational aspect of those prohibitions on copying today, and the strange violence that accompanies the enforcement of intellectual-property law: the raids on American working-class and immigrant neighborhoods where counterfeits are sold, the involvement of mafias and gangsters, the sporadic global "war on copying" undertaken by the United States and other governments, with their discourses of moral hygiene (the protection of "economic health"). As with many other taboos, such as those on sexual practices, on killing people, or on transformation, the exceptions to the prohibition on copying are numerous. Taboos manage the boundaries between acceptable and unacceptable forms of human behaviors, particularly those concerned with basic framings of what it means to be human.

Girard argues that all taboos relate to mimesis, since all taboos exist to head off the possibility of mimetic violence. Insofar as taboos are an exclusively human phenomenon, taboos are necessary because the mimetic capacity of humans is much greater than that of animals, and there is thus also the possibility of a violence that could destroy the whole community. Taboos on incest, hostility to twins, suspicion of actors and the theater—these are ways of curtailing a perceived threat from excessively mimetic phenomena and the danger that is/was thought to accompany them.

But if, as I have argued, violence is only a single way of framing mimesis, it cannot be used as the basis of an explanation of all the prohibitions on mimetic activity. Thus, we are still left with the question: Why should there be so many taboos on mimesis? Canetti's tentative response regarding the reason for there being so many taboos on transformation, one that I think can be broadly applied to mimesis, is that man "felt as though there was nothing but movement everywhere and that his own being was in a state of continual flux; and this inevitably aroused in him a desire for solidity and permanence only to be satisfied through prohibitions on transformations." This state of "continual flux" would be something similar to the state out of which the transformation of things emerges in the Chuang Tzu quote at the beginning of this chapter. This fear of flux would also be connected with the fear of the feminine which I discussed in relation to Irigaray, which manifests as "anxiety and hence efforts to name and designate causes."[23] Finally, this fear is a fear of nonduality, of nondifference.

Body Worlds: The Copying of the Dead

Bataille, the most important theorist of taboo, relates taboos to death, and to the fear of the contagiousness of death. Arguably, the first "copies" that human beings encountered in the current pejorative sense of the word were the dead bodies of members of their community. These dead bodies were "undifferentiated," in the sense that Girard uses the word—emptied of the life and agency that gave them a particular form, returning back to nature and formlessness. Hamlet's speech in the graveyard, as he holds the jester Yorick's skull in his hand, says it all in this regard. Dead bodies already have some of the qualities attributed to copies, in the sense that they are viewed as degraded versions of originals; and the taboo-like atmosphere that surrounds copying and copies may come from a feeling of discomfort, even horror, with dead bodies, in which being no longer manifests itself in outward appearance. Reduced to skeletons, most human beings look the same; reduced to matter, the human body is absorbed back into the earth. Since this is the fate not only of the objects and beings around us, but of our selves as well, it is a matter of both fear and fascination. It is understandable that we all have fears of the contagiousness of death, because of the real possibility of infection with microorganisms—but, more deeply, because the "same" death that we see before us also exists within our own bodies as a potential which, in time, will certainly become a reality and thus "spread" to us. It is for this reason that yogis practicing Hindu or Buddhist tantra in various Himalayan and South Asian cultures meditate in the cremation grounds. They seek to confront, open themselves to this contagion, work with it. In the words of one of the workers at the Manikarnika cremation ghat at Varanasi, "Burning is learning . . . cremation is education!"

The anatomist Gunther von Hagens explores this tension between sameness and uniqueness in his *Body Worlds* exhibitions, which feature a series of "sculptures" based on dissection and manipulation of dead human bodies which have been "plastinated" (Hagens' term for his patented method of embalming bodies so that their tissue structure remains intact and solid). Hagens' work walks a number of fine lines, including those separating science and art, inside and outside, commerce and knowledge. When I took my students to visit the exhibition at the Science Center in Toronto, we could not agree on whether the sculptures could be called "copies." Some students angrily refused to see that the show had anything to do with copying—a response that I am inclined to interpret as marking the proximity of the taboo. I myself was nauseated by the show, and felt the contagious presence of the dead in that otherwise antiseptic and dutifully labeled museum environment.

Certainly, those "sculptures" are not living human beings; and because of the process by which they are preserved, but also transformed to make them accessible to manipulation, they are not "dead bodies" in the strict sense of the term, either. Most of the marks of individuality, of outward appearance, such as skin, eyes, and so on, have been removed, leaving a more or less generic human body, which at the same time is to be used in exemplary ways (to demonstrate a certain network of tissues, to show a pathology such as lung cancer, but also to show certain athletic postures). Hagens amplifies the questions concerning what these bodies are by giving each "sculpture" a name ("The Dancer" etc.) and signing each of the sculptures with his own name, displayed on a metal plaque next to the object. One walks the exhibition halls in a strange mimetic fog, recognizing with discomfort in each exposed fleshy, sinewy, or skeletal structure the mimetic reflection of one's own body. At the same time, one is confronted by the modes of mimetic framing (the science project, the philosophical questions, the sculpture, etc.) which Hagens uses to channel this energy—all of which feel slightly fake, as though they were mere alibis for something more vicarious, more obscene in the sense of taboo breaking.

The mimetic energy in *Body Worlds* is contagious. Hagens' work has been imitated by a number of "copycat" exhibitions with titles like *Bodies, The Amazing Human Body, Body Exploration, Bodies Re-*

vealed, Mysteries of the Human Body, and *The Universe Within.* According to the *Body Worlds* website, these exhibitions appropriate and imitate Hagens' techniques and have "plagiarized the unique expressive character of many of his distinctive plastinate specimens."[24] In addition, they obtain dead bodies from questionable sources, using unclaimed or abandoned corpses. Hagens' own solicitation for body donors is prominently displayed both in his exhibitions and on his website, where we learn that he has already signed up 6,800 donors, of whom 350 are deceased. One senses the contagion here—the urge to become either a donor or a fellow plastinator, each a way of mastering or appearing to master death.

The Right to Transform and the Right to Copy

If Hagens' work points to the continuing existence of a taboo on copying, it also points to a shift that we are undergoing in relation to the forces that constitute the taboo. It may be, as Girard argues, that we are no longer able to believe in the myths that hold up the taboos, and are forced to face directly the mimetic mechanisms that the myths and taboos managed and/or obscured. Nevertheless, these structures still are able to exert considerable force, especially insofar as they are currently being brought to a point of crisis that threatens their existence. Or perhaps their existence is as strong as ever, only now transposed to the marketplace and the economy, where the desacralized taboo takes the form of intellectual-property law, and the use of copyright, patents, and trademarks to control mimetic transformations. These laws are backed up by the omnipresent codes and passwords which function as ritualized protectors of identities in places where transformation is rapid, such as the banking system, the airport, the supermarket checkout, and the Internet.

Although the most aggressive defenders of copyright law have done their best to link copyright breach to terrorism, gang violence, drugs, and other scapegoats, the extravagance of such claims, even when on occasion they are actually well-founded, reveals the functioning of the taboo. But such desperate exaggeration, which accompanies the widespread production and exchange of "copies," regardless of aggressively enforced laws concerning copyright infringement, still suggests the diminished power of this taboo. The fact that the means of producing copies are increasingly available to larger and larger groups of people around the world instead reveals the taboo that protects and naturalizes capitalist production modes, in particular the myth of the naturalness of the commodity and of private property.

Would it be possible to speak of something like a right to transform, or even a right to copy? The philosophical topic of rights is a vast one that is beyond the scope of this book. Let us note here that Hegel based his philosophy of right (German: *Recht*, also meaning law), a foundational text for modern political theory, around a person who is constituted through his or her right to claim certain things as property, such as his or her body. Copyright is a matter of rights, and of course property. But so are more fundamental questions, including whether I have the right to my own image (if someone photographs me) or whether I can call my shadow my own.[25]

The closest that the UN's Universal Declaration of Human Rights comes to a statement concerning copying is Article 27, section 2, which affirms the moral and economic basis of intellectual-property law, stating that "everyone has the right to the protection of the moral and material interests resulting from any scientific, literary, or artistic production of which he is the author." The document repeatedly affirms consent and choice and even the fact that we are "born free," but does so for the most part in relation to preexisting structures that are said to be the basis of society: the family, the law, the nation, work—not to mention the hegemony of the English language (in which the document is written), the assumed universality of the pronoun "he," and the assumption that literature, science, and art are supposedly universal human activities.

The most interesting article in terms of our topic is Article 13: "1. Everyone has the right to freedom of movement and residence within the borders of each state. 2. Everyone has the right to leave any country, including his own, and to return to his country." The intensity of our fear of transformation is revealed if we remove the phrase "within the borders of each state" from section 1, or change section 2 to read "the right to enter any country." The right of migration, of asylum, of movement between nations, of transformation of national or ethnic identity and affiliation is a mimetic issue akin to those concerning Chuang Tzu and the butterfly, or Woody Allen's Zelig. The boundaries, borders, institutions that currently restrict such movement are themselves mimetically constructed entities (think, for example, of that remarkable twentieth-century artifact, the passport) that strongly resist transformation, not to mention those who fake their identities, who do not belong, who belong elsewhere—and thus have the right to "return to [their] country"! Parallel to this would be the right not to be coercively transformed—as in the case of indigenous peoples around the world who are violently inserted into political and economic regimes that they do not want to be assimilated to, or made "the same" as. A rethinking of mimesis could support a politics that also established the right to nonequivalence.

Gender politics are increasingly centered on the right to transform, and any examination of this right has to find its way through and in the categories of Man and Woman. Patrick Califia has set out very well the complexity of the possible identifications related to sexuality and gender that someone might make in the course of a lifetime, and his/her argument concerns the right to make those identifications.[26] Gender dysphoria is another excellent example of the Chuang Tzu / butterfly situation and the decision/question/problem of how to identify, which, again, is a mimetic problem—or a problem of how we think about mimesis, and about the difference between saying "like a man" and "as a man." The right to multiple "coming outs," which serve as moments of self-identification, the situational dynamics of "giving an account of oneself," to use Judith Butler's phrase, emerge out of respect for the shifting processes by which the chaotic and creative flux of transformation is given name and form.[27]

The right to transform would also play a determining role in political issues around life and death—for example, abortion, euthanasia, even the boundaries and limits around exhibitions like Hagens' *Body Worlds* or Gregor Schneider's recent proposal for gallery exhibitions of dying or dead people. It also arises in an increasing number of biomedical crises, such as the use of performance-enhancing drugs in sports (does one have the right to artificially transform oneself to increase one's abilities?), the outer limits of cosmetic or plastic surgery (do I have the right to transfigure my entire body?), the taboo on certain psychoactive drugs as agents of various kinds of transformation (what Timothy Leary called the fifth freedom, ecstasy as psychic self-transformation). More generally still, there are the enormously challenging bioethical debates that surround the genetic modification of organisms, from seeds to animals to human beings. Should one be allowed to transform the basic biological structure of organisms? Should one be allowed to claim any such innovations as private property? Should those identified as genetically different from prevailing norms be accorded the same rights?

Every one of these situations is highly complex, and I can do little here to address that complexity. But let us affirm that the right to copy, and to transform ourselves and our environment through copying, is a political issue in ways that go far beyond intellectual-property law. It is hard to see what is at stake in each of these situations—where a crisis occurs because a particular transformation is being coerced or denied, where the right to transform self and other is accorded to one group and denied to another, with no recognition of the way that a particular understanding and framing of mimesis is being deployed. By "framing" I mean taboos, laws, discourses, and so on. Such framings, which are eminently ideological but which are presented as natural, manipulate our fears of the remarkable plasticity of mimesis; they set standards for what is called "original" and what a "copy," what is "real" and what is "fake," who belongs and who is an imposter, what is fixed and what is allowed to change, what is called "natural" or "unnatural."

If we understand that the many crises of transformation outlined above have a certain commonality—fear of mimetic transformation, and a reliance on taboo-like structures and framings to manage this fear—very basic political questions arise. What if we faced our fear? Could we do without such framings entirely? To what degree can we even speak of "rights" when thinking about the inexorable processes of transformation by which we and everything around us are constituted as entities? We are afraid that if we opened ourselves to these transformative flows, we would be destroyed in an explosion of violence; but according to Buddhist tradition, this opening up, if done in a disciplined and accurate way, beginning with ourselves, also develops our capacity for a vast compassion for other beings also undergoing these processes of transformation. And this compassion—not conceptual, but developed through practice as experience and realization—is surely the basis of any future politics.

4/Copying as Deception

A Hundred Thousand Harry Potters

"Chinese Market Awash in Fake Potter Books," reads a recent *New York Times* headline. The article goes on to describe the proliferation of unauthorized Harry Potter books in China, in the days leading up to the publication of the seventh book in the series. As the author of the article says, these fake books are "copious." It is worth quoting the description of the books in full:

> There are the books, like the phony seventh novel, that masquerade as works written by Ms. Rowling. There are the copies of the genuine items, in both English and Chinese, scanned, reprinted, bound and sold for a fraction of the authorized texts.
>
> As in some other countries, there are the unauthorized translations of real Harry Potter books, as well as books published under the imprint of major Chinese publishing houses, about which the publishers themselves say they have no knowledge. And there are

> the novels by budding Chinese writers hoping to piggyback on the success of the series—sometimes only to have their fake Potters copied by underground publishers who, naturally, pay them no royalties.[1]

Copia is here in all its variety and diversity: there is the original series of texts, copyrighted by J. K. Rowling, copies of which have been published and sold by various publishers worldwide, in accordance with existing intellectual-property laws; there are identical copies of the originals, made and sold by other people; there are works that appear outwardly to be originals but that contain something different; there are translations of the originals which may or may not bear a resemblance to the originals; there are new works that anticipate the additions to the present series that J. K. Rowling may or may not write in the future, some written in the style of Rowling's originals, others appropriating names and characters from other texts and placing them in new settings (titles cited include *Harry Potter and the Half-Blooded Relative Prince, Harry Potter and the Hiking Dragon,* and *Harry Potter and the Chinese Empire*); and there is the proliferation of copies—fake books themselves being copied and sold. If one were to enumerate all the varieties of copying of Harry Potter books that are possible or actually in existence, as in Jorge Luis Borges' fantasia "The Library of Babel," they would approximately equal the sum total of creative acts possible in the universe. In India the publication of *Harry Potter in Calcutta,* in Russia the exploits of the orphaned girl and aspiring wizard Tanya Grotter, and in Belarus the fictional hero Porri Gatter—all indicate a global proliferation of Potter copies of various kinds.[2]

The official discourse concerning copying as a debased or immoral activity is here in full, too: "fake," "phony," "masquerade," "genuine," "unauthorized," "real," "pure invention," "borrow," "lifting," and so on. The "underground" publishers "naturally" pay no royalties, presumably because they are the inherently evil scapegoats of the piece, thwarting the equally natural aspirations to fame and fortune of those naïve writers hoping to "piggyback" on Rowling's success. All of these words serve to communicate the notion that copying is wrong because it is an act of deception and that, in the words of a lawyer representing J. K. Rowling's literary agency, "some of these examples seem to suggest that J. K. Rowling actually wrote the books. . . . It is possible that people might buy those believing them to be part of the series, and obviously they'd be disappointed."

Deception

One of the principal arguments made against copying is that it involves an act of deception. Something is presented in the guise of something else. This something is produced so that its outward appearance corresponds to something else, to something that it is not. Plato's moral objections to mimesis follow from this point: the imitations of the artist, he says, confuse the viewer as to what is real, and what the source of realness actually is (though Plato's dialogues themselves, which have a more or less fictional structure, are also imitations in this sense). And Plato is not alone in making this objection—most philosophical examinations of copying, even the most radical, split copying into two forms, one good and one bad, the bad one associated with deception. Intellectual-property law today also invokes the notion that producers should have the right to be identified with their work, and that consumers should be protected from acts of deception (a.k.a. fraud) where an inauthentic, inferior copy is passed off as an authentic original. Even the most pro-copyleft person, suspicious of the way this argument has been used to justify the corporate takeover of the public domain, recognizes that selling medicines that do not contain the drugs they claim to have is wrong. Most of us accept that J. K. Rowling and her publishers similarly have a right to profit from their inventions in a way that others do not—whatever limits we set on that right.

But to cast this proliferating new Chinese Harry Potter literature in exclusively negative terms hardly does it justice. American legal scholar and judge Richard Posner gives the following definition of plagiarism: "A judgment of plagiarism requires that the copying, besides being deceitful in the sense of misleading the intended readers, induce reliance by them. By this I mean that the reader does something because he thinks the plagiarizing work original that he would not have done had he known the truth."[3]

If plagiarism means appropriating someone else's work and presenting it as one's own, then the same criteria could be applied to plagiarism's inversion, forgery, in which one presents one's own work as the work of someone else. There are many situations where work we do is presented as being by someone else, but not all "induce reliance." As for the bootlegged editions of actually existing official translations of Harry Potter books—no doubt they do. New translations of existing works might or might not induce reliance—the case is arguable either way. Works of varying degrees of novelty, even if attributed to J. K. Rowling, certainly involve some degree of deception; but given the highly variable historical uses of the name of the author, the deception here might not induce reliance. The use of the author's name might be funny, meaningless, or irrelevant (not to say irreverent) to the audience for these books. Isn't the proliferation of texts as in the Harry Potter case characteristic of any genre, with its variations on a set of conventions? We see it, for example, in the massive network of fan fiction sites on the Internet, some of which also appropriate Harry Potter into their own set of narrative structures.[4] Such appropriations have a long history in literature—from oral folk traditions, where local embellishments enhance a shared repertoire of stories and songs, to Renaissance theater, where Shakespeare and Marlowe continually lifted plots, characters, and dialogue for their own works. The complexity of the many possible economic and social arrangements of textual dissemination is reduced by contemporary intellectual-property law to a situation of legal ownership and consumer's rights—despite the necessarily copious quality of any textual communication or event.

Who are the "intended readers" in this case? Those intended by J. K. Rowling? Or the ones envisaged by the plagiarizers? What if the audience consists of those too poor to afford full-price books? When copying increasingly involves situations in which a cultural product is taken from one milieu, culture, or community and reproduced in another, with an inevitable recontextualization, isn't some form of deception or misrecognition inevitable? As Ted Striphas points out, the appearance of the Potter fakes is a direct consequence of the coercive installation of a particular kind of Western economic and legal framework around the world.[5] Such an "intentional" installation inevitably leads to the "unintentional" transfiguration of the imported cultural forms—yet this of course is what global IP law seeks to forestall. In what way is a bootlegged J. K. Rowling book published in China different from an iPod manufactured in a Chinese factory and sold in the United States, or an Indian call center, with its Joshes, Sams, and Sues, making bookings for American motel chains? Aren't these all situations in which there is a slippage in the identification and presentation of things? And how are we to "know the truth" in such situations? Via a tag that says "Made in China"? Or a detailed report on the labor practices of a particular factory that manufactures goods? What if we reached a point where we were able to make a "judgment of plagiarism" or fraud (not to say deception) against the totality of our current global social-political-economic apparatus, which, as we know from Marx, is predicated on obscuring the reality of the labor processes and conditions that go into making a commodity?

The transnational context of the story is significant, for any situation or event of translation—such as a book written in England being presented in Chinese—inevitably involves an appropriation of and re-presentation of the original; all involved in the networks of dissemination (author, publisher, translators, retailers, readers, journalists) participate in deception when the complexity of this transformation is elided. This deception is most pernicious when it is authorized and when the translation is presented "as" the original—as though the translated copy were not a transformation of the original, and in some sense and to some degree a new work. But this is the form that we are most comfortable encountering it in—which is to say that at some level we want to be deceived, and we want to believe that the copy is exactly the same as the original, even though, for all the reasons set out in the previous chapters, it cannot be. Furthermore, the "original" itself is also necessarily an appropriation, translation, imitation of other materials now presented, packaged, and marketed in ways that objectively constitute deception.

Deception has been a fundamental philosophical problem, from Plato through Descartes, Hegel, Nietzsche, and the poststructuralists.[6] "Do not lie" is the first example Kant gives of a categorical imperative—a maxim one follows in order that it become a universal law.[7] The argument that copying is wrong because it is deceptive rests on the belief that it is always possible to name and describe things correctly, to say what an original is, and for things to present themselves correctly via their outward appearance. As we have already seen, in the absence of any locatable essence, all production involves the presentation of something in the guise of something else, and the possibility, in effect, of deception. In what follows, I am not arguing that deception is always a good thing, or that no punishment should be meted out to those who deceive people by selling them poorly made copies of things that lack the qualities of the original that the buyer might expect. There are values I recognize that make it very difficult to defend deception, let alone to write "in praise of copying." For example: the ability to give an accurate account of oneself or some thing, both to oneself and others; recognition of and striving for situationally valuable originality; respect for the contributions and the integrity of others (beings, things, systems).[8] But, as I will show, the problem of deception complicates any attempt to speak of authenticity, to affirm originals in opposition to copies, or to speak of any kind of grounding, whether ethical, political, scientific, or otherwise.

Le Musée de la Contrefaçon

The Museum of Counterfeiting occupies a large Belle Epoque building in the west of Paris. It is owned by the Union des Fabricants, an industry organization that focuses on piracy, bootlegging, forgery, and counterfeiting, particularly of the kinds of luxury goods that Paris is famous for. It consists of a series of rooms with exhibition cases and texts—much like any other museum. The museum as a whole is a visual display, and, for the most part, what it displays are copies of outward appearance, which, in our consumer culture, usually means packaging: bottles of L'Oréal and Hugo Boss perfumes, Cointreau and other liqueurs, even bottles of Mr. Clean! Beyond that, what is striking is that most of the counterfeited products also rely heavily on a distinctive but easily copied outer appearance. They are products that are in some sense shells: for example, handbags, which are structures of surface that contain and conceal an interior; watches, which have a "face" that reveals or manifests time, and which for the most part hide the mechanism that tracks time; Peugeot car bonnets, lamps, and mirrors, which all constitute parts of the car's outward appearance. Even the clothes on display—Adidas T-shirts, sneakers of various brands, Lacoste shirts, and Versace suits—are surfaces that we put on as shell-like outward appearances.

Amid these objects, whose outer appearance, for better or for worse, can be easily imitated and forged, the digital objects sit strangely. There is a display of DVDs—a boxed set of episodes from the TV show *Friends,* as well as disks of *Pocahontas* and other Disney movies; a tiny exhibit of audio cassettes and CDs (Johnny Halliday and Abba), already looking as though they came from a prehistoric era; and two cases of computer software, Adobe and Microsoft products, complete with packaging.

But the relation between outward appearance and essence, which structures our fantasies about what objects and/or products are, breaks down in the digital world, where the question of packaging is more or less irrelevant. No MP3s are on display, because there is nothing there to display: one could show a computer screen (the node through which most of our fascination with appearance passes in the digital age) with a copy of iTunes running, but the digital counterfeit would look identical to the original. One could show an MP3 player, but the process of miniaturization means that, again, there would be little to see in an iPod Nano. In the contemporary world of counterfeiting and forgery, the easily discovered and confiscated copies of Microsoft Office and Adobe products which used to be on sale in computer shopping malls in Hong Kong or on street corners in New York City have all but disappeared in favor of digitally downloaded "cracked" software and "warez" (pirated software distributed online), whose only moment of visibility or presence is a piece of spam e-mail or a listing on a search engine. The packaging of computer software is an almost entirely abstract phenomenon, like a piece of conceptual art. There is no particular reason a software package should be a certain size; it could be as small as the smallest material storage product you can find—let's say, a microSD flash memory card—or as large as a washing machine or a fifty-inch plasma TV, which cost about the same as some pieces of software.

If we are speaking of deception, then the entire world of product packaging and marketing is an act of deception. But few people are interested in living in a world of amorphously colored miscellaneous liquids (carbonated beverages, cleaning products, fermented beverages, perfumes), twisted and forged metals (watches, car bonnets, cellphone shells, belt buckles), that have neither name nor recognizable form. The Museum of Counterfeiting, which itself is housed in a handsome old mansion in one of the wealthiest parts of Paris, recognizes this. There must be a spectacle, something to see—otherwise, no one would bother to visit the museum. These displays of suboptimal forgeries and equally suboptimal originals (apparently, no one wanted to donate anything too valuable!) have the quality of a minor Pop Art exhibition at a provincial art school, long after the heyday of Pop itself. Warhol and company had a genuine appreciation for the beauties of advertising and packaging, and they amplified these qualities by changing the scale of the packaging and displaying the packages in art galleries. The museum in Paris cannot or will not consciously glamorize counterfeiting, because if the forgeries were too persuasive, or too creative, the line between the copy and the original would start to erode, and one would start to see the officially produced "originals" instead as particular types or styles of copy. And something like this does happen with a slightly cheesy but fun pink Christian Dior handbag that actually looks better in its counterfeited, fake-leather, poorly stitched copy version, redolent of Mike Kelley and other appropriation artists.

We've already established the problematic nature of assertions of authenticity when it comes to industrially produced items like Louis Vuitton handbags, which cite a tradition of craftsmanship and of unique artisanal fabrication, or a very limited seriality, while flourishing in the age of global capital. The many ways in which companies seek to protect their legal ownership of their products, and their exclusive right to produce them, all paradoxically involve strategies of representation that make their products more vulnerable to being copied, imitated, faked, or forged. Brand creation and maintenance relies on the inscription of the brand as a set of signs onto an object that finally and inevitably evades that inscription; and philosophically and otherwise, such an object is empty of essence. The pathos of deception, and the supposed need to protect the public from its harmful effects, are used to enable corporations and wealthy individuals to legally enforce their right to extract maximum profits from a given situation.

Fakes and Forgeries

Forgery has a long and venerable history in material culture generally, and in the visual arts and literature in particular. Anthony Grafton tracks the historical record of forgery back to the Middle Kingdom of ancient Egypt; he describes the proliferation of "dubious orations and plays" in fourth-century B.C. Athens, and the presence of scholars in Greece and Rome who could distinguish fakes and originals: "Of the 130 plays of Plautus in circulation, the scholar Varro judged 109 to be forged and 21 genuine."[9] As both Grafton and Sándor Radnóti show, even a formal definition of what constitutes forgery requires a clear understanding of a complex set of historical distinctions and developments. Notably, not all fakes are forgeries: the mistaken identity of many fakes is the result of a variety of erroneous attributions that accrue over time. Radnóti describes the complex ecology of what are called "Rembrandt paintings." Thanks to the work of the Rembrandt Research Project, "Rembrandt" now seems to splinter into dozens of entities: "students, members of his workshop and people from his surroundings, painters under his influence, the juvenilia of later masters, etc."[10] Copying was an integral part of the visual arts until the eighteenth century, when the rise of originality and authenticity as aesthetic values, and the rise of various forms of intellectual-property law, retrospectively transformed the copier into a forger, and the multiplicity of similars and imitations into fakes. Where copying persisted, in name if not in fact, it was relegated to the applied arts or to folk arts, until the postmodern period, when the pervasiveness of copying in industrial societies was recognized.

According to Radnóti, the nineteenth century was the golden age of forgery, "on account of that broad and blurred continuum in which restoration, renovation, new copies of old objects, old objects assembled into new ones, and the creative use of retrospective, historicizing fantasy to produce new objects were all lumped together" (91). Thus, the transformation of copying parallels a more general shift in our relation to history: to Darwin and evolution as the theory of the emergence of present organisms from a flow of historical similars and dissimilars; to Freud and the origins of a current symptom in a repetition of a past trauma; to Hegel and Marx's dialectical view of history as a chain of responses and counter-responses. The fetishism of authenticity, of history—so elegantly parodied in *Bouvard and Pécuchet* by that great nineteenth-century forger Flaubert—is accompanied by the progressive undermining of the self-evident status of present subjects and objects. Claims of authenticity attract a swarm of inauthentic duplicates and simulacra, just as for Gilles Deleuze the actual image emerges out of the cloud of virtual similars and dissimilars. The profusion of the luxurious, the unique, the exceedingly rare or precious object, be it Thomas Jefferson's wine collection, Picasso's art, a precious stone, or a nearly extinct animal, draws along with it a swarm of forged copies, all of which have at various times deceived those who claim to be able to distinguish them.[11] Forgery is contagious. Grafton gives a number of examples—including Erasmus—of great experts who were devoted to unmasking fakes, and who also produced forgeries themselves.

As for forgery defined *strictu sensu* as the deliberate act of passing off your own work as the work of another: according to Radnóti, there are two strong inducements to deliberate forgery, "money and competition" (5). Both are evident with the global fake Harry Potter books, yet only some of these books are labeled as being written by "J. K. Rowling." (And what of that most predictable strategy of the forger, the slightly misspelled name—"K. J. Rowling," "J. K. Rawling," etc.? Is this really an attempt to deceive, while marking off a difference? Would it make any difference in a legal case?) Forgery brings us back to Girard's theory of mimetic desire—for the copy exists only because there is competition or rivalry for something, and one way of satisfying the desire of those who are competing is to ensure that there are multiples of the desired object. Indeed, one of the most common claims about how to recognize a fake is that it responds too directly to the fashions and styles of the moment in which it was produced; and that, over time, the historicity of this desire separates itself from the historicity of the desire that went into the production of the original, even though both original and copy share a number of characteristics. (I will address the validity of this argument later.)

As Anthony Grafton argues, knowledge and deception have a strangely dialectical intimacy: knowledge evolves as a way of overcoming the deceptive claims of the forger, while the skill of the forger grows in response to the increasingly subtle determinations of the expert or connoisseur. Something similar occurs in Hegel, where the dialectic consists in a progressive overcoming of strata of self-deception; and in Lacanian psychology, where the patient's acts of deception and self-deception are manifested in the analytic session as transference, which, when taken as the object of reflection in analysis, is what makes knowledge and self-knowledge possible. Indeed, for Lacan, "les non-dupes errent" ("the non-duped make an error"), since deception is constitutive of the symbolic order (a.k.a. "les noms du père," "the names of the father") and our necessary insertion into it. The forger understands the symbolic order only too well—but what is the status of forgeries in the Real? The question is an enormously complicated one, since, as Žižek points out, "by 'pretending to be something,' by 'acting as if we were something,' we assume a certain place in the intersubjective symbolic network, and it is this external place that defines our true position."[12] He gives the example of Alicia and Devlin in Alfred Hitchcock's film *Notorious,* who fake being lovers in order to distract attention from their mission, but who actually do love each other, or become lovers by faking it. But this does not always happen, and the paths between the symbolic order and the Real remain labyrinthine and full of dead ends.

Warfare and Camouflage

> All warfare is based on deception. Therefore, when capable, feign incapacity; when active, inactivity. When near, make it appear that you are far away; when far away, that you are near.
>
> —Sun Tzu, *The Art of War,* trans. Samuel Griffith

The association between warfare and copying is a venerable one. Sun Tzu, author of one of the earliest known treatises on war, considered deception, based on the manipulation of appearances, to be essential: one should try to remain invisible as long as possible, so that one's enemies do not know how or where to deploy their forces; at the same time, one tries to get the enemy forces to reveal themselves so that one always knows where they are. One thinks also of the Trojan horse in the wars of Troy, and the use of stealth, the object that is not what it seems to be, to penetrate the defenses of the enemy. This should amaze us—that against the application of the greatest force, the copy might have a power. It should also indicate some of the potential dangers in copying.

A 1946 British Royal Air Force (RAF) film entitled *Visual Deception* divides the topic into two categories: disguise and display. Thus, one camouflages where one is, while giving the appearance that one is where one is actually not. Camouflage involves blending into the environment, disappearing into one's surroundings. Display involves convincing the enemy forces that they should direct their resources against a place where one is not, and involves not only the simulation of the presence of people, machines, buildings, etc. with false ones, but the simulation of the movement of one's forces, the simulation of a response to an attacking enemy, and the simulation of a successful attack by the enemy (the setting of fires in empty places so that it looks as if the enemy has successfully struck a target). One might simulate a whole environment, so that the entire terrain of battle becomes indeterminate. This, then, raises the question of how to stop one's own forces from confusing the simulation with the reality (trying to land on a simulated landing strip or trying to make their way home using a false path)—the same issue that Roger Caillois raised in his classic essay on mimetic phenomena, "Mimicry and Legendary Psychasthenia," in which he points out that the camouflaged insect runs the risk of being eaten by other animals who mistake the insect for the leaf that it is pretending to be.[13] The RAF used the word "starfish" to describe the configuration of decoys around an actual target to be protected. But if the arrangement of the decoys is too predictable, then it no longer functions as camouflage.

One can speak of the seductiveness of appearance here: by attacking a perceived target, the aggressive party commits itself to a decision concerning appearances, concerning where things are and what they are. The camouflage team invites the aggressors to make this commitment, thereby revealing themselves and laying themselves open to attack. A copy, as described in the film *Visual Deception,* is merely the hollow shell of an object—outward appearance pared to the absolute minimum required to persuade aggressors of its reality. This need not be the case. A fisherman might use a feather or a plastic worm on a hook to catch a fish, or might use a real fly or a worm. Similarly, an army might use real troops to invite an attack. In either case, the object is not what it appears. The real fish contains a hook. The Trojan horse contains soldiers. The abandoned village is surrounded by real forces. As Dudley Clarke, leader of Britain's deception unit in World War II said, one surrounds the lie with truth.

According to Paul Virilio, the practice of stealth by the U.S. military in the First Gulf War and subsequent battles—involving planes, weapons, and personnel that are prepared in advance to avoid detection systems, and technologies that scramble the capabilities of such systems—was more significant than the sheer quantity of force on the battlefield. Conversely, the ability of the Iraqis to withstand the increasingly "precise" and overwhelming use of force that accompanied the American technologies of stealth was linked to the multiplication of false targets, whether these were buildings hidden underground beyond detection, or personnel, such as the various doppelgängers of Saddam Hussein and his family that were said to exist.[14]

Stealth and camouflage are of course only two examples of the much broader practice of deception in conflict—strategies which would also have to include the use of the spy and the double agent, of surveillance operations (from the bugged telephone to the "washing" of the Internet), and of propaganda as a way to manufacture the consent of allied populations and the dissent of the enemy. Iraq's fictional weapons of mass destruction, presented by the U.S. and British governments as a compelling reason for invasion, are a recent example. Virilio argues that the use of such strategies of deception is part of an increasing militarization of civil society, where methods (such as deception) that were once limited to formally declared warfare are now applied continually as basic tools of managing civil society. The claims that there is a "War on Drugs" or a "War on Terrorism" serve today to provide moral justification for the continued use of deception as a political tool of government.

Power, Survival, and Competition

In his essay on mimicry, which I touched on above, Caillois recounts a vast array of cases of mimicry within the animal kingdom, only to systematically discount any biological advantage of such phenomena. Caillois brilliantly and perhaps arbitrarily neutralizes the problem of deception by arguing that the similarity between an insect's outer appearance and its environment confers no particular benefit—indeed, it is a "luxury," for the insect that looks like a twig may avoid being recognized by predators, but is then eaten by those who eat twigs.[15] Stripped of purpose, mimesis is a matter of succumbing to the lure of the environment, to the recognition that "nature is everywhere the same." In a perverse way, Caillois affirms nondual "sameness" as the necessary condition for the appearance of mimetic phenomena. Distinction may be an illusion, but it is pervasive; and it is pursued with an incredible intensity, because life and death appear to be at stake. It is here that the tension between biological materialism and Buddhism reaches its highest point, where the disseminative chains of biological survival, operating through sexual reproduction and environmental appropriation, fold into inevitable impermanence, chance, and the dissolving of purposiveness in ultimate nonduality.

The purpose of deception and the production of the copy whose purpose is to deceive, in many of the cases that we have looked at, is to ensure survival, or to engage in competition. In an elegant meditation on the possibility of a "history of the lie," Derrida points out that the lie is often connected to action, specifically political action.[16] Dudley Clarke formalized this in World War II by saying that the goal of the Allies' deception was to change not what the enemy thought, but what the enemy did, and to commit the enemy to acting or not acting in a particular way. Such a concept of action remains strangely indifferent to questions of truth or deception, originality or copying. The forger of religious manuscripts seeks to affirm a particular dogma to colleagues or peers. The forger of paintings seeks to persuade a buyer that he possesses a hitherto unknown Rembrandt or Matisse, and to sell the painting without anyone's discovering the truth. Deceptive action aims at a temporary advantage, which may end with exposure, but which may also be sustained by further lies. The copy rests within this web of deception, yet it is equally available to those who pursue the truth.

Copying and deception, then, are connected to power, which (to repeat Russell's definition) is "the production of intended effects."[17] The power of the copy lies with this ability to produce an action—in Posner's words, to "induce reliance" deceptively. Thus the use of the deceptive copy in warfare; in nature, where survival against predators is a constant concern; in politics, where the goal is to maintain or to take hold of power; and in capitalist economics today, where the goal is to persuade people that there is no viable alternative to their continued participation in this particular political/economic structure.

Such uses of the deceptive power of the copy are embodied in the rhetoric of the leader—the statues and images, buildings and ceremonies, that concretize power, install, and legitimate it by appearing to fix its form and ownership. But, as with the Harry Potter books, the meaning of "deception" in this case is complex; and it is here, too, that copying is revealed as a power of the poor, of folk cultures, and what Derrida calls the possibility of civil disobedience. In "The Storyteller," Walter Benjamin says:

> The fairy tale tells us of the earliest arrangements that mankind made to shake off the nightmare which myth had placed upon its chest. In the figure of the fool it shows us how mankind "acts dumb" toward myth; in the figure of the youngest brother, it

> shows us how one's chances increase as the mythical primordial time is left behind; in the figure of the youth who sets out to learn what fear is, it shows us that the things we are afraid of can be seen through; in the figure of the wiseacre, it shows us that the questions posed by myth are simple-minded, like the riddle of the Sphinx; in the shape of the animals which come to the aid of the child in the fairy tale, it shows that nature not only is subservient to myth, but much prefers to be aligned with man. The wisest thing—so the fairy tale taught mankind in olden times, and teaches children to this day—is to meet the forces of the mythical world with cunning and with high spirits.[18]

Thus the importance of the ruses of the protagonist in Jaroslav Hašek's novel *The Good Soldier Schweik* (whose protagonist avoids fighting in World War I through his cunningly literal obedience to the orders of his superiors), and of deception in guerrilla warfare and popular resistance movements. Deception functions in such situations because the superior knowledge of terrain, environment, and community of those invaded or colonized allows them to make distinctions that are invisible or imperceptible to outsiders who control terrain through superior force. But deception is also a motif in folk songs, in hip-hop and rock lyrics, in the culture of carnival—in the enigmatic lyrics of Michael Jackson's "Billie Jean," for example, with their multiple dramas of misrecognition and misconception, and the interminable theme of jealous love.

But the copy also complicates Russell's definition of "power," since the deceptiveness of the copy suggests the ability to produce unintended effects as well; Shakespeare's comedies and the drama of the imposter are evidence in this regard. The deceptive copy, which may or may not produce an intended effect, is no less powerful either way. Yet it still is subject to impermanence. Radnóti argues that the forger engages in an attack on the modern faith in the historicity of the art object—because the forger, working in a particular time and place, aims at producing an object that unmistakably belongs to a different time and place. But according to Grafton:

> If any law holds for all forgery, it is quite simply that any forger, however deft, imprints the pattern and texture of his own period's life, thought and language on the past he hopes to make seem real and vivid. But the very details he deploys, however deeply they impress his immediate public, will eventually make his trickery stand out in bold relief, when they are observed by later readers who will recognize the forger's period superimposed on the forgery's. Nothing becomes obsolete like a period vision of an older period.[19]

How do we reconcile Radnóti's and Grafton's apparently antithetical positions? Grafton's view of forgery assumes it is inevitable that all forgeries will be revealed as such over time—but proving this would require the ability to accurately label every case of deception as such, and this would be impossible, even if it constituted the faith of a classical historian. Doubtless, it is true that some forgeries are revealed in this way—although, as Grafton notes, subsequent evaluations of the exposure of a particular forgery might later reveal the designation of "forgery" to be erroneous. But the notion that the forger might be even partly successful in his/her refutation of the historicity of the object is disturbing for obvious reasons, and suggests one of the more radical dimensions of emptiness, properly understood: that temporal and physical determinations are themselves nothing more than traces without any grounding essence. More provocatively, a certain "sameness," beyond all concepts, reveals itself when the historicity of the object is carefully examined. This is the forger's ultimate resource—but it is equally the resource of anyone who claims to produce authentic originals, or to write "true histories."

Buzz Rickson's

Historically, one of the measures of masterful craftsmanship has been the maker's ability to deceive other experts or masters as to the authenticity of an object. Thus, we learn that when the great Chinese calligrapher Wang Xizhi wrote a letter to Emperor Mu, the emperor had a calligrapher make a copy of the letter, and sent it back to Wang with a response. "At first, Xizhi did not recognize [that it was a copy]. He examined it more closely, then sighed and said: 'This fellow almost confounded the real.'"[20] Here, the ability to deceive is a measure of skill. And the most profound, not to say perplexing, kind of deception is self-deception: in this case, the possibility that the copy made by another is so accurate that Xizhi mistakes it for what he should be most familiar with, namely his own work.

But there are other master forgers to whom the ability to deceive is a matter of indifference. Buzz Rickson's in Japan, celebrated by William Gibson in his novel *Pattern Recognition* (2003), is a company that makes replicas of classic American military clothing. There is no attempt to persuade the potential client that these jackets are "real," and surely most people who buy this clothing, which is by no means cheap, consider it more prestigious to be wearing a Buzz Rickson's replica than they do an original flying jacket. Buzz Rickson's is proud of its attention to detail. To quote the company's website regarding its American A-2 flying jacket:

> Our simple desire has been to perfectly revive the vintage A-2 with all of its nuances, a challenge we have succeeded at through an extensive research and development process encompassing

> many experiments on materials, component parts, patterns and sewing fabrication. Leather is, perhaps, the most important aspect of the A-2 to recreate correctly, thus we were very particular in the selection process of the hides; specifically, we have expended more time and energy in the study of different tanning processes than any other aspect of the A-2, with experiments in this area being repeated constantly.[21]

The website goes on to detail the company's experiments on tanning, which go beyond the reverse engineering of a product being made today, to encompass questions regarding the history of the materials used, the techniques used to produce the originals, and the work-arounds required to produce a highly accurate replica of the original as it appears today.

Canadian pianist and composer Glenn Gould, notorious for his disavowal of the apparently authentic pleasures of concert performance in favor of the satisfaction of studio recordings, argued that "the role of the forger, of the unknown maker of unauthenticated goods, is emblematic of electronic culture. And when the forger is done honor for his craft, and no longer reviled for his acquisitiveness, the arts will have become a truly integral part of our civilization."[22]

For Gould, studio technology allowed for the fabrication of much more complex and precise musical experiences than are possible in even the most virtuosic concert performance. In his view, studio recordings deceive the ear that seeks in them an accurate representation of a live performance, which is itself a representation of a piece of sheet music. But given all the questions we have raised about authenticity, "accurate representation" cannot be the primary consideration when it comes to establishing the value or even the identity of the object—especially at a time when the technological possibilities for fabricating sensuous experiences of all kinds are proliferating rapidly, when there may be a number of different routes that can be taken in producing a certain thing or feeling or perception. More than that, there might be radically different hypotheses, interpretations, performances, arrangements, or manifestations emerging from the "same" original—as anyone listening to Gould's startlingly different early and late versions of Bach's *Goldberg Variations* can hear. And knowledge of the decisions and methods that go into a particular mode of fabrication is something that could be part of the pleasure of enjoying an object, as is the case with Buzz Rickson's. Or these decisions could be something that is hidden.

Universal Deception and Play

Another word for the distortion that Plato so dislikes in mimesis is "play." In his writings on mimesis, Walter Benjamin attaches extraordinary importance to play, going so far as to claim that semblance plus play equals imitation.[23] In the inauthenticity which Plato ascribes to the copy, we can already find a hint of play. It is this idea of play which Aristotle picks up on in his *Poetics*, his famous rejoinder to Plato, concerning the value of mimesis. Through imitation as it occurs in art, in the games of children, in performative play, we learn. We also take pleasure in imitation, in the act of recognizing correspondences and making them—a pleasure that is surely found today in the countless acts of digital replication. While Plato sees in imitation a threat to the republic, organizations like the Recording Industry Association of America see the end of business or an act of criminal self-interest in the copies being passed around on the Internet. But everyone has experienced the joy of copying and the way it opens up to us the mysteries of play.

In play, the line between appearance and reality is blurred and objectivity gives way to fascination—the intense concentration on a fabricated situation experienced by the theatergoer, the video-gamer, or the soccer fan. Play—according to Johan Huizinga and Roger Caillois, its two great theorists—is a circumscribed activity; but this is hardly a universal belief. From the point of view of a variety of Asian religious traditions including certain forms of Taoism, Vedanta, and Buddhism, the great deception consists precisely in thinking that things are what they seem to be—that subject and object have essences. The word *Lila* in Sanskrit signifies "divine play," and the activity of the gods in causing events to unfold as they do. There is nothing higher than play, and submission to this play is mimetic: it means allowing oneself to become a manifestation of the will of the gods. This is to say that there is a charm to being deceived, and that this charm is what holds us to the realm of appearances. Of course, at other times, it hurts. Having said this, we should note that there is a major fault line passing through many of these traditions as regards the status of play. For example, the great nineteenth-century Bengali saint Ramakrishna, a tantric devotee of Kali and celebrator of the "funhouse" of appearance, was presented in the twentieth century by his followers as the advocate of Advaita Vedanta and the reduction of the phenomenal world to illusion.[24]

The belief that the world itself is a fake, an illusion, or a deception is a surprisingly pervasive one—probably because there is some truth to it. It takes various forms in various Asian religions and cultures, and it also appears in a particularly dramatic form in early Christian Gnosticism, where the entire universe is said to be the abortive creation of a malign deity; we now find ourselves trapped in it, and must find our way out. Gnosticism manifests in a variety of contemporary situations, including the novels of Philip K. Dick, the hit movie *The Matrix,* and Jean Baudrillard's popular text *Simulations,* all of which assume that everyday life is a mere copy of some other reality, which awaits revelation.[25] In Žižek's Lacanian reading, these are classic fantasies of the Big Other—that projection of an agent that embodies the Real lurking behind the façade of the Symbolic.

But "simulation" is only one aspect of Baudrillard's thought, one that in his lesser-known later work he counterposes with what he calls a theory of seduction, which is the culmination of his profound meditations on symbolic exchange and his critique of the political economy of the sign. Baudrillard defines seduction in terms of "play, challenges, duels, the strategy of appearances." Seduction, to his mind, is specifically a practice of folk cultures, which he sets in opposition to the order of global capital: "We are living today in nonsense, and if simulation is its disenchanted form, seduction is its enchanted form. Anatomy is not destiny, nor is politics: seduction is destiny. It is what remains of a magical, fateful world, a risky vertiginous and predestined world; it is what is quietly effective in a visibly efficient and stolid world." Furthermore: "The strategy of seduction is one of deception. It lies in wait for all that tends to confuse itself with its reality. And it is potentially a source of fabulous strength. For if production can only produce objects or real signs, and thereby obtain some power—seduction, by producing only illusions, obtains all powers, including the power to return production and reality to their fundamental illusion."[26]

We need to adjust Baudrillard a little here, since—assuming Baudrillard is not an adherent of Advaita Vedanta—"fundamental illusion" is a nihilistic formulation of a rather crude postmodern kind. But if instead we say "fundamental nonduality," what Baudrillard writes is in accord with those cultures that I have already described as practicing a non-Platonic form of copying: traditional, indigenous, and/or subaltern cultures of various kinds; the avant-gardes; contemporary subcultures; suppressed forms of the feminine. All are trivialized when described in terms of make-believe, superstition, false consciousness, nostalgia, and self-deception, but all can also be affirmed in terms of a play that is sovereign, without guarantees (such as Kant's categorical imperative not to deceive), and eminently functional.

Baudrillard observes: "The game's sole principle, though it is never posed as universal, is that by choosing the rule one is delivered from the law. Without a psychological or metaphysical foundation, the rule has no grounding in belief. One neither believes nor disbelieves a rule—one observes it" (133). This helps us to understand the implacable opposition between folk cultures and intellectual-property law. Folk practices of copying are based on rules—the open secrets of the symbolic order that one does not have to believe in but that one still has to observe. One is not deceived when one practices magic, for example, since magic "is a ritual for the maintenance of the world as a play of analogical relations, a cyclical progression where everything is linked together by their signs. An immense game, rule governs magic" (139). And it is through the rule that the power of copia is unfolded.

The Zone of Appearance

Orson Welles's film *F for Fake* begins by exploring the ways in which even the most distinguished art market experts can be fooled by a master art forger, such as Elmyr de Hory, who can crank out a passable Matisse before lunch, or Clifford Irving, who begins as a biographer of de Hory and then creates his own forgery, a fake "as told to" biography of reclusive millionaire Howard Hughes, which is accepted by the American mass media as being true. *F for Fake* was filmed in Ibiza, in a "high-society" milieu, and the camera is extraordinarily attentive to the little gestures of style, attitude, and irony which allow the individuals involved to compose themselves and their stories, simultaneously revealing and concealing themselves, making a display of their apparent cunning. The contagiousness of mimesis is evident everywhere in the film, in which Welles creates a cinematic hoax of his own, identifying himself with a long history of tricksters, flim-flam artists, and hustlers who play with the seductiveness of image, story, and name to seduce us into believing that a fabrication is genuine.

The realm of appearance is a political realm—this is something that we already know. Elmyr de Hory, star of *F for Fake,* knows it—and we would probably say that he exploits this fact, playing with the appropriations made by art experts and galleries, manipulating the magic they bestow on objects in declaring them authentic, catching them at their most vulnerable point, in the object itself, and the repetition of traces, surfaces, and styles. Hory delights us with the skill he shows in making copies, the humor and spectacle and sexiness of his parties on Ibiza, the teasing game of appearances, high-society clothing, people passing themselves off as minor royalty, some of them from the correct background, some of them not (all of them performing a part). We also recognize that when de Hory sells a Modigliani to a gallery and it turns out to be forged, he faces and deserves arrest, trial, and punishment for making false claims about the painting and for profiting from these false claims.

Welles himself knows it better than anyone, for the cinema is a zone of appearance par excellence, where what is allowed to appear manifests itself in order to seduce, to fascinate, causing us to temporarily forget the illusory nature of light and shadow projected onto a wall or screen. Thus the marvelous sequence in *F for Fake* of men filmed watching a mini-skirted woman walking down the street, not knowing that they themselves are being filmed. The sequence is obviously somewhat hokey, despite the beauty of the model and the sincerity of her male admirers. It feels like bait, and it is, throwing us off guard, so that when Welles promises that there will be nothing but truth for the next hour, we take what he says at face value—"the next hour" apparently meaning for the duration of the film. When that hour, but not the film, is up, it is not a shot of a girl's legs or the horde of men watching her that distracts us from realizing that truth time is over, but a very solemn homage to the timeless, monumental beauty of the cathedral at Chartres, which one sucker on YouTube describes as "the most profound moment in the history of the cinema." It is this moment of profundity and seriousness that makes the ensuing bogus yarn regarding Picasso's seduction by Welles's muse Oja Kodar feel so persuasive and so real.

Something appears to be what it is not. Thus, we fall in love, we make commitments, we live in the shadow of potential error. This possibility is encoded in different ways in all the great religions: as the Native American trickster; as Maya, illusion; as Mara the tempter; as the devil, the great imitator. And those who trade in such deception are considered evil. Many of our vices, our pleasures too, are connected with copying: pornography and drugs are both regarded with ambivalence because they are thought to be the imitation of something real and precious, and for this reason they are said to deceive, even when they are consumed by a conscious and rational subject. So much of our laughter and suffering is the result of deception and its effects, and the inevitable yet unexpected slippages in how things appear to be at different times.

As we saw in Chapter 1, the definition of "copy" shifts radically according to whether we think of it in terms of originals and imitations, or in terms of repeated essenceless forms. What if the realm of outward appearance is understood as a realm of fabrication and play, rather than as a crudely construed realm of "things as they are" and the various distressing letdowns that come from the expectation that things should be what they seem? The assertion, made in Mahayana Buddhism (but also in other South Asian religious traditions), that the realm of appearance is the realm of *samsara*, of illusion, implies a pervasive, omnipresent field of deception—not just the deception of the inauthentic or imitative object that tries to seduce us into mistaking it for an original, but the deception constituted by our insistent belief that the phenomenal world is composed of originals at all. The great deception is the belief that things have a continued and fixed essence, the mindset that ignores the impermanence and the continuous transformation of phenomena.

François Jullien argues that classical Chinese culture understood emptiness in a particular way, one that differed from "Indo-European" and "Buddhist" notions.

> This implies a fundamental difference in the status of the invisible in Greece and in China. What is invisible in a Greek model-form *(eidos)* belongs to the order of the intelligible, the "mind's eye," or theory. Meanwhile, the kind of invisible that interests the Chinese is that which is not yet visible in the undifferentiated basis of all, way upstream from any process. The intervening stages of "the subtle" and "the infinitesimal" *(wei)* make the transition possible, and it is on these that the sage/general relies to orientate himself.[27]

In terms of warfare, this means tricking one's enemies into revealing themselves while one remains invisible oneself until gaining the advantage. In terms of sickness, it emphasizes prevention over cure. Jullien argues that this practice emerged out of a concern with efficacy: while things are "upstream" and have not yet occurred, it is easy to influence them through minimal action; but when they are actually happening, even a major mobilization of force can have only a minor influence. Heidegger, himself a keen although mostly secret scholar of East Asian philosophy, adopted a similar strategy, emphasizing the importance of revealing and concealing, emergence and withdrawal in the coming-to-presence of Being, without ever directly using the word "emptiness." Heidegger's celebrated critique of phenomenology could easily be described as an expounding of the paradoxes of a "phenomenology of deception," and his failures in this regard point to what Lacoue-Labarthe describes as a secret but pervasive mimetology.[28]

Today, it's commonplace to say that structure controls appearance. The realm of deception today isn't the realm of appearance per se; it involves control over what appears at a much more basic level—genetic modification, nanotechnology, biomimicry, and surveillance beginning with the most elemental forms of matter. The same with identity, which today is concerned less with names and faces than with chemical modifications, neuro-enhancers, and so on—mimetic deception achieved by the manipulation of appearance through its substrates. Geophysical warfare, control over weather; the computer, the use of code, the Internet—all of these in various ways instantiate Platonic idealism, and the determination of reality through instrumental models. Nevertheless, despite the apparent superiority of structure to appearance, the passion with which scientists continue to attempt to manipulate appearance and surface is quite striking. The beauty and terror of the life-world, its radical contingency, may well be subject to technical mastery (or ascetic withdrawal, for that matter), but if there were no value to appearance *qua* appearance, there would be no reason to intervene in it at all.

The Copy as Scapegoat

Who is to blame when we are deceived? Can there even *be* deception unless there is someone to blame? Deception occurs in the realm of action, as does copying, since the copy is likewise said to be something which is produced, rather than something which is found in nature. Even the term "self-deception" suggests a split: the self contains an other who can be blamed for deception. According to Girard, the act of finding someone to blame for all of those mimetic tricks, slips, transformations, all those copies, and then punishing that someone, is what holds society together. The escalation of mimetic energies and rivalry, and the sparks of mimetic violence that occur as a result, would threaten to engulf the whole world if there were no possibility of focusing all the collective violence onto a scapegoat figure—someone who is seen as responsible for the trouble, who can then be sacrificed in a ritual that brings a temporary sense of unity and togetherness to the community.

While Girard says that this victim is arbitrary, Lacoue-Labarthe points out that often the figure is already associated with mimetic activity—he or she is an actor, an imposter, a copyright breaker, or one of the many mythical trickster figures who incarnate the energies of mimetic deception: Khezr in Islam; Legba in Haitian Vodoun; Hermes in Greek mythology; Mara the tempter in Buddhism; Satan in Christianity . . . More broadly, we can say that the word "copy" today carries with it that negative judgment, that subtle but decisive abjection from the realm of legitimacy, that indicates scapegoating. The "copy" is the scapegoat for the immense and apparently unsolvable problems that mimesis, as a basic constituent of our situation, poses for us. It allows us to imagine that there are things called "copies" that can be identified, fixed as such, judged, and punished or removed entirely from our existence, so that we can live in a world where everything is what it appears to be, where deception never occurs, and where no one is ever deceived by anybody or anything—or at least where deception is always recognizable and manageable.

Girard believes that it is possible for us as humans to stop deceiving ourselves regarding mimesis, and to take responsibility for our own deluded outlooks. He links this process of becoming undeceived to a Christian narrative of salvation, Christ being the example who refuses and reverses the logic of sacrificial scapegoating. Compassion can be cultivated, reversing the otherwise inevitable slide toward mimetic violence. Girard has changed his position over the years: he no longer sees this reversal as being an exclusively Christian insight, but now recognizes that many religions have made this observation and drawn consequences from it, although he insists that only Christianity reveals the entirety of the mechanism of sacrificial victimage.[29] In Buddhist terms, compassion emerges from a recognition of the interdependence of all phenomena. If one is undeceived, mimesis can be recognized as the freeplay of traces and phenomena, and one can participate in this play in an undeluded way, recognizing the unstable, impermanent nature of things, participating in it as necessary. But if one were undeceived in this sense, there would be no need to renounce copying, or to scapegoat or disavow it, or even to fix it as having a particular essence. All attempts at transcending mimesis—including the Christian one—necessarily reinstate it.

There is a key Mahayana Buddhist practice called, in Tibetan, *Lojong* ("mind training"), which consists of a series of slogans, or, to use Baudrillard's word, "rules," to be memorized and practiced.[30] One of these slogans is: "Put all blame on the one." Basically, this means that when one encounters adverse conditions, one should put the blame on oneself, rather than on other people or factors. This is done not as a guilt trip, but as a way of cutting through the chains of cause and effect known as "karma," which one participates in through blaming and through retaliating, because one mistakes an illusion for a reality. It is a practice whose purpose is not to turn one into, in the words of one of my teachers, a "doormat," but to allow one to see through the dense network of deception *(samsara)* that the self ultimately creates through ignorance about how things are. The potential deceptions of mimesis are not a problem in Lojong practice, since every phenomenon is viewed as a copy that one can engage with through the observation of a rule, and thus the practice of copying can itself be appropriated for the training of the mind. The slogan after "Put all blame on the one" reads: "Meditate on everyone as kind." This is arguably a projection, but it entails a practice that consists of unraveling another projection—that of hostile separateness—and generating compassion based on understanding the reality of the other's situation, and one's own. One "copies" the thoughts of an enlightened mind in order to gradually "train" or "practice" (both of these words implying a repetition) and thereby become what one at first "merely" imitates. In Lacanian terms, the praxis of Lojong consists in the way one treats the real via the symbolic; and thus, despite Žižek's excoriations, Lacanian thought and Buddhism find an affinity.[31]

We have already observed the importance of action in relation to deception and the copy. In many Buddhist traditions, the word for action would be "karma," meaning not merely what happens to you or what you do, but the aggregated set of causes, effects, actions, and consequences that produce phenomenal reality as it appears to us and as we experience it. It is thus connected to "mimetic desire," and more generally to the anthropology of jealousy, envy, and other emotions, which are believed in many traditional cultures to play an important role in what happens to a person. We are left with the following radical proposition: appearance, which irreducibly involves some kind of subjective element or frame, is the coming together of karmas, jealousies, desires, and so on. If there is to be a scapegoat, an entity blamed for the power of mimesis, this should be not some exterior force, but our selves, our egos. "Put all blame on the one." No individual or collective act of liberation can begin without recognizing this.

Turnitin.com

I have been grading forged or copied papers ever since I started teaching. In the late 1990s, when the World Wide Web was new, it was easy to spot a paper that had been downloaded from the *New Yorker*'s website, or copied manually from the magazine itself. At other times, a radical disjuncture between parts of a paper—some very sophisticated, others illiterate—gave the game away. Today, in a world of online term paper archives, of "factories" that write customized papers and accept credit cards, discerning the authenticity of a paper has become more difficult, so that I can no longer be 100 percent sure when I am reading original work or a copy, when I am reading an honest expression of the student's understanding of a topic or when I am being deceived. My university uses a service called Turnitin.com, an online archive of texts and essays, to which a student or a professor may submit a paper and which will send, in response, a readout of the paper's originality or lack thereof. I have not used this service, in part because it doesn't seem to work very well, in part because I find the idea of this kind of surveillance offensive. I believe that the writing and grading of a paper is a contract between myself and a student, and that it is my responsibility to determine the quality of the work. If, as a professor, the assignments I give are mere copies of other assignments, then I should not be surprised if the work I receive is also a copy—for, in a sense, a copy is what I have asked for. From the point of view of Warholian or Duchampian aesthetics, the ability to select an appropriate text for a paper and present it as your own work, or even the ability to hire the right person to write a quality paper for you, could be seen as admirable qualities. This is particularly the case in the business world that students are likely to find themselves in upon graduating—a world in which, despite exhortations to "think outside the box," they will also need to master what's "in the box," much of which consists of the ability to manipulate copies. I was impressed when my colleague Kenneth Goldsmith, in his class on "uncreative writing" at the University of Pennsylvania, gave students an assignment instructing them to download a paper online for a project. Perhaps the best response to this assignment would be for a student to actually write the paper and then hand it in as if it had been downloaded.

Having said this, I do believe there is such a thing as deception in writing papers, and that it matters. I feel an almost visceral sense of displeasure when I discover a plagiarized paper, partly because I know all the work that will go into documenting the plagiarism, meeting with the student, and negotiating the various administrative protocols that come into play in such a situation; and partly because I am offended by the student's refusal to think, whether that thought be the recognition of another's thought or the manifestation of the student's own creativity. It is hostility to this impoverishment that governs much of my own morality regarding copying. Someone who copies merely to make money, or to get a degree, or to get a job as a professor, offends me. If, added to this, the copying is shoddy and attempts to pass off inferior materials or craftsmanship as the real thing, again in order to make money, then that seems wrong to me as well. Someone who copies out of love, out of a desire to share or a desire for knowledge, out of fascination with the magics of production and form, seems to me in a different category.

But I would like to dwell a little longer on this "visceral sense of displeasure," which has a whiff of the taboo on copying that I discussed in Chapter 3. In *Counterfeit Money,* Derrida questions whether we are mistaken, deceived, when we think of gifts as being an object or a thing, and as relating to possession, to "having" rather than "being." He argues that what is important is not the object called the "gift" but the event of giving and receiving, and the states produced by that event. When a student "hands in" a plagiarized paper to be "marked," and I "give" it a grade, aren't we initiating something like an event of giving and receiving, of exchange? And isn't my visceral sense of displeasure related to the fact that this event is disturbed, and that the ontological categories of student, teacher, and text are thereby confused? The objection to the "copy," in the pejorative sense of the word, would then not only be that it is different from what it appears to be, and thus deceives, but that it triggers or catalyzes an event in which everything involved is perturbed and becomes questionable: the status of student, teacher, and assignment.

But if this happens so easily, isn't it because those entities are not as stable as they appear to be? The stability of the system, such as it is, requires a certain amount of deception, a certain exercise of power—something like the public secret (mentioned by Taussig), which everyone is aware of but no one can talk about. It requires violence in order to keep functioning, to ensure that everyone continues to look and act like the entity they're supposed to be. After all, overt plagiarism is only the crudest form of misrepresentation of identity in the classroom. What of the student or professor who takes neuro-enhancers in order to write a paper? Should he or she be subject to the same restrictions imposed on a Tour de France cyclist, concerning chemically enhanced performance? Is he or she merely mimicking competence? And what of collaboration with a smart friend, or advice from a knowledgeable parent? How far away is that from hiring PhilosophyWriters.com to "advise" you by writing a "model" paper for you?

Only rarely, in fact, does someone hand in a paper that is a 100 percent duplicate of another piece of work. This may be the real scandal of Goldsmith's assignment: he did not say "Add a new intro," "Take it and change it until it becomes something new," or "Be creative with it." And this scandal also seems to be felt by students who, lazy as they are, feel obliged to leave some trace of their own minds on the appropriations they wish to pass off as their own work. They do this not only because they want to throw the Googling professor off the scent, but because at some level they also feel a need to transform the text, to mark it and make it their own, even if this means defacing an otherwise accomplished text. Goldsmith's most successful works as a writer/artist/poet are those in which any embellishments of the original material are minor: his book *Day,* for example, consists of a transcript of every word contained in a single edition of the *New York Times*—stock quotes and advertisements included—formatted and published as a book. *The Weather* transcribes a year of weather forecasts on a New York radio station. Goldsmith reveals the deception while carefully concealing his own originality, which consists in small but essential decisions as to format, scale, name, and medium. When the composer Morton Feldman told Karlheinz Stockhausen that his secret lay in never manipulating the sounds, Stockhausen shrewdly replied, "Not even a little bit?"[32]

I am still struggling over what to do about my plagiarizing students, with their passionate devotion to copying, and the institutional requirements that force them to hide their copying and deceive their teachers, or reveal it and be accused of a lack of originality, or go through a ritualized, laborious performance of originality that bears little resemblance to the standards they have to meet in everyday life. In this light, I cannot see the deceptions of my plagiarizing students as an entirely bad thing. No doubt, it is symptomatic of the decadence of contemporary society, where the ability to produce a certain appearance situationally is more valuable than the slow development of substantive skills. The lazy plagiarism, executed in a sloppy way, without attention or care, is easily discovered, and is on a par with a lazy piece of original work, or a lazy assignment. But good deception, and the copy that it fabricates, can possess profound insight, not only into situational requirements, such as what the teacher is looking for, but also into the object being copied. For the great forgers and fakers—the reverse engineers, the modelers, the fabricators—are often, as we have seen, very attentive to the object. In this sense, broadening the repertoire of means of deceiving is not a goal entirely unworthy of an educator.

5/Montage

Cornell's Boxes

So far, we've been concerned with copies that more or less involve a whole imitating a whole. It's true that Elias Canetti's observations on the different degrees of transformation already suggest a difference between a mere surface or superficial imitation and a total interior and exterior transformation. But even in these situations, one assumes the presence of a whole. Canetti's donkey dressed in a lion skin is either taken to be a lion or revealed in fact to be a donkey. Kafka's ape says he has no choice but to consider himself a man now. According to the film *Divine Horsemen,* when the *loa* take over worshipers in voodoo rites, they do so completely. Objectively, of course, the old body is there at least. I am concerned less with the objective truth of the situation than with the belief—the valuation of imitation as involving "complete states." "A copy" is usually a copy of *something.*

In Joseph Cornell's assemblage *A Parrot for Juan Gris* (1954), a parrot painted onto a cut piece of wood sits perched in a wooden box, as if in a birdcage. Several metal rings hang from the box; its sides are covered with newspaper clippings and pieces of old maps, which are superimposed so that they obscure and interrupt each other. On the floor of the box is a white plastic ball. Is Cornell's box a copy? Not if we think of copying as being concerned with a whole imitating a whole (though, interestingly, Cornell made at least four different versions of *A Parrot for Juan Gris,* using different clippings and objects but producing a recognizably similar form). Yet, assembled out of fragments or samples of mass-produced objects, the box does imitate a parrot in a cage, and it works because of the way it leverages many of the peculiar powers we have already examined that constitute copying as creative act.

Cornell's boxes reappear in William Gibson's cyberpunk novel *Count Zero* as an early metaphor for the radical disjunctures of cyberspace. In the world of computers, "cut and paste" is a dominant metaphor; more broadly, fragmentation, pastiche, and juxtaposition are characteristic of postmodernity. Indeed, art critic Nicolas Bourriaud has claimed that montage, and other practices of citation, repetition, and appropriation, constitute the core of a contemporary art practice which he variously names "relational aesthetics" and "postproduction."[1] Bourriaud situates this centrality of montage within the context of globalization, the culture of the DJ as curator, selecter, and sequencer of a vast historical and geographical archive, and the Internet as a limitless virtual space of assemblages governed by the logic of the click and the hypertextual trace. Montage also plays a key role in contemporary critical theories, from Derrida's theory of the trace, to Deleuze and Guattari's assemblages and disjunctive syntheses, to Badiou's ontology of the pure multiple and Latour's actor networks. In summary, a logic of montage is pervasive today, and as soon as we speak of acts of imitation without essence, this logic asserts itself as the construction of the similar from the dissimilar. If copying means the presencing of the *eidos,* or outward appearance, in a place where it does not belong, that disjuncture that constitutes making a copy is always already an act of montage. Where there are no essences, how could something like montage not be operative? But what, then, are the politics of montage today? Montage, throughout the twentieth century, was seen as an oppositional or emancipatory avant-garde practice. But isn't it today, to paraphrase Žižek, a crucial ideological determinant of late capitalism in its multiple-choice, build-to-suit, do-it-your-way™ mode?[2]

A word on terminology here. The assemblage of a new artifact from fragments of preexisting objects or forms is one of the key practices of modernist aesthetics, and can be dated back as far as 1869 when Lautréamont proclaimed the beauty of "the chance meeting, on a dissecting table, of a sewing machine and an umbrella." The word "collage" (from the French, meaning "pasting," "gluing") is used to describe two-dimensional artworks incorporating found objects, while three-dimensional visual artworks made in this way, such as Cornell's boxes, are often known as "assemblages." The word "montage" (from the French, meaning "mounting") was originally used by Soviet filmmakers such as Sergei Eisenstein in the 1920s to describe the process of editing and assembling film footage—but it was quickly taken up by Berlin Dadaists such as Hannah Höch for their "photomontages." Today, "montage" and "collage" are often used interchangeably, especially to describe works of assemblage in other media, such as sound or text. What is a montage? We can define it as involving the following elements:

1. Fragmentation of the original object, or copies of it;
2. Tactile exploration of material (the use of scissors, "cut and paste," etc.);
3. Juxtaposition with other objects (combinatory methods);

4. Selection of a particular combination of elements or fragments;
5. Naming and framing of the new object.

Let's look at each of these elements in more detail.[3]

Parts and Wholes

Montage implies that a whole has been broken, even if it is then reassembled into a new whole. Something is broken in a montage, and in most successful montages you can still see the break, which is often what makes them funny. A montage in which you really cannot tell that two unrelated sources have been cut and pasted together may be successful in all kinds of ways (Douglas Kahn talks about the time it takes to recognize the truth behind a montage's hoax of being real, and we might also mention "pastiche"),[4] but it is unlikely to be funny. Until you see the break. A montage involves an act of destruction. Hannah Höch—who helped to invent montage as a politically driven art project with the Berlin Dadaists, during and after World War I—titled her most famous piece *Cut with the Kitchen Knife Dada through the Last Weimar Beer-Belly Epoch of Germany.*

Anthropologist Stanley Tambiah, in his discussion of James George Frazer's theories of magic, sees metonymy as a key part of magic—the substitution of one thing for another with which it is closely associated.[5] And montage is evidence of this: the artist makes a copy from fragments of other objects, or from wholes that now become components of a newer, larger assemblage. But these things are not just fragmentary bits to be reassembled like Lego bricks; they in some sense contain the whole from which they come—and when they are placed in a montage, the transformation of the fragments that occurs also exerts an effect on the original from which the fragments came. For the Situationists, the power of *détournement,* the transformation of preexisting elements in a new ensemble, "stems from the double meaning, from the enrichment of most of the terms by the coexistence within them of their old and new senses."[6] It is in this sense that montage is a practice of copying, since it often involves the citation of the old object in the new.

The fragments of the materials are contagious; insofar as the likeness of the original can still be perceived in them, they also have semblance, similarity. Thus, for Eisenstein, the first principle of montage is the "associatively infectious capabilities" of the actor to be filmed.[7] The montage radiates back out into the world from which it was cut—and the radiation takes in not only the actual original that the fragments were taken from, but all those "like" it.

Tactile Exploration of Materials

Having broken something up—whether it's an image, a tape, a text, or (in film) an event—there follows what Kahn calls the "tactile exploration of materials."[8] You push them around, juxtapose them in various ways, in a variety of combinations. Something similar to Taussig's contagious magic, a basis of the tactile aspect of mimesis, is involved here. You could call this a gleeful pleasure in destruction or creation, where—rather than being the passive consumer of an object that was bought whole and retains instructions for its maintenance as an approved preexisting whole (I am thinking of warranties for computers that are considered broken as soon as you show evidence of tampering with the integrity of the outer shell)—one gets to perform magic on the object, rearrange it at will.

This is by no means an activity limited to the avant-garde. Think of examples of mosaic in public or private spaces—Gaudí's parks and buildings in Barcelona, the Watts tower in Los Angeles, Nek Chand's rock garden in Chandigarh—all of which make use of arrangements of fragmented, recycled found objects. Or the accumulations of texts and other material traces in 9/11 commemorative shrines and in anti-Iraq war protests. Folk cultures around the world and throughout history have likewise enjoyed montage as an activity. For example, quilting, when done in groups, entails collaboratively making a quilt out of fragments of old cloth, mixing and stitching them into a new pattern. Montage is obviously important for cultures that can't afford to buy new things—it is a poor people's art. We see this in hip-hop too, where it was in part economics that led DJs like Grandmaster Flash and Afrika Bambaataa to assemble new dance tracks out of fragments of cheap old vinyl records. Again, think of the importance of tactility, the hand, as the DJ scratches records and manipulates turntables and mixers—or the collective handiwork of quilting bees. This is not just a matter of pragmatic hands-on fabrication techniques. The touch of the *monteur* (DJ or quilter) sends a shiver through matter, marks it temporarily as the *monteur*'s own, asserts a kind of freedom with it and a claim to the right to transform it. Just as the touch of a lover asserts that right.

Combination and Selection

We are therefore talking not just about the wisdom of the hand, but also about the invocation of some other order of powers in manipulating and playing with reality. What are those powers? For if things do not have a preexisting order, what is it that causes a particular combination of things to occur, a particular sequence of events to unfold? These are difficult questions. In contemporary Western cultures with an explicitly named practice of montage, the outcomes are often attributed to "chance," "randomness," "spontaneity," and other formulations of a materialist indeterminacy. The Surrealists believed the Freudian unconscious was the source of the spontaneous orderings of their "automatic writing." In traditional societies, the forces that produce a particular configuration of elements could equally be articulated through various operative kinds of cosmic determinism or, to use Baudrillard's word, predestination—astrology, for example, or divine forces. In the translations from traditional to modern, from East to West, a particular ritual technique for revealing an invisible but deterministic order—for example, the text/divination scheme of the *I Ching*—becomes a technique for exploring randomness: "synchronicity" for Carl Jung, "indeterminacy" for John Cage. Similarly, Jack Kerouac's "spontaneous bop poetics" translated Zen and bop-era jazz into an ethics of creative spontaneity.

The tension between determinism and random ordering can be found throughout the modernist practice of montage. In the case of the politically motivated montages of German Dada or early Soviet cinema, the artists are seeking a particular rational, cognitive, and/or emotional meaning. Eisenstein was scornful of attempts to use montage as a way of either producing a faux realism or manifesting some kind of "cosmic" order; he argued forcefully that the goal of cinema should be to influence the audience, and that such influence should be at the service of the revolution. He and filmmaker Dziga Vertov both also speak of a "science" of choosing cuts, edits, and combinations. Yet it's unclear whether the actual organization of Soviet cinema or German Dada was finally determined by such considerations. Eisenstein dismissed what he called "mimesis of form" in favor of "mimesis of principle"; but the nature of this principle, which lurked behind appearances and which organized film montage, remained obscure, and Eisenstein was forced to use metaphors such as "bone structure"—which strike an appropriately materialist tone, but explain very little.[9] In his list of the elements of montage, the final element for Eisenstein was an "obfuscation of the schema" that was linked to a personalized rhythm. Vertov likewise speaks of a "rhythmical order" which determines the final organization of materials. And no doubt this would be true for other time-based arts, such as music or poetry.[10]

The choice of a particular combination and arrangement of materials ends the flux which the act of cutting, permutating, and combining initiates. But this is not necessarily a permanent decision, and the advent of process over product, itself a result of the development of various technologies that facilitate montage, means that elements can circulate more continuously, as in a kaleidoscope. Thus, the DJ, who is sometimes known as the "selecter," mixes records live and performs, and each performance will be in some way different from the others, even if the records are the same. The computer also allows an almost infinite iterability of materials—endless edits and remixes.

Name and Frame

> The title is 50 percent of the work.
>
> —Jack Smith, interview with Sylvère Lotringer, *Semiotext(e)*, 3, no. 2 (1978)

Finally, the framing of the montage is important. In the case of Cornell's boxes, we can speak of a literal frame around the objects. Whatever processes by which the discontinuous or unconnected objects involved in the montage are transformed into a new unique object can be considered part of the framing. In mail art, the frame might be the envelope in which the art is mailed. Or it might be the reproduction of the montage—photocopy, cassette tape, MP3 file—that is generated. Or in certain cases, the frame might be the particular use that is made of the montage—the act of giving the montage to someone, of situating it in a particular place.

Consider the twelve-inch single released in 1983 by Grandmaster Flash entitled "The Adventures of Flash on the Wheels of Steel." The title recalls Lautréamont and Hannah Höch in its expansive brashness. Although built around a montage of rhythm tracks taken from Chic, Queen, and others, cut together with noisy rupturing scratches, there are recognizable words in the track too—mostly the names of the MCs, and Flash himself, who signs his mix by cutting his own name rhythmically into the mix. But the brief sections of rapping on the track are also full of names—in particular, star signs and brand names (of jeans), the two historical book ends of Walter Benjamin's mimetic faculty. Marketing and branding, which have both been appropriated by hip-hop, consist partly in giving a name to an industrially produced commodity. Generic bottles and cans of fabricated sugar and water, themselves "a montage of attractions," are given the name "Coca-Cola," for example—transforming them into the desire- and fantasy-charged objects that appear in our stores.

Hip-hop is extraordinarily concerned with naming. Rammellzee speaks of the act of painting over other people's graffiti tags (their names written on train cars, walls, or fences) as an act of assassination. Hip-hop rhymes often consist of little more than a listing of names, and, as with graf writers, inventing a name is an important part of hip-hop's performativity. One of contemporary popular music's most important tropes, the incorporation of a group of individuals into a band with a name ("The Beatles," "The White Stripes") is itself a manifestation of the power of naming—no doubt one that draws on Afrodiasporic cultural history and the erasure of African names by the slave trade.

In the world of quilting, too, naming is a way of framing a pattern. One pattern may have several different names. Thus, the pattern "Indian Trail" is known also as "Winding Walk, Rambling Road, Old Maid's Ramble, Storm at Sea, Flying Dutchman, North Wind, Weather Vane, Climbing Rose and a few more." And the names have a contagious power: "If a young person slept under a quilt whose name might have an adverse effect on the character, the name was changed. One never put a young boy under a Wandering Foot, lest he turn out to be a wanderer, and the quilt would be called instead Turkey Tracks."[11]

If a montage is made of fragments appropriated from another object or objects, naming the montage entifies it—actualizes it, gives it coherence as a discrete entity. Thus the importance of intertitles in silent cinema, as a way of organizing an otherwise murky montage of film clips. There is power in naming, because naming brings together the heterogeneous energies of various fragments and unifies them in a particular name/form. There are many possible names in many languages for an object, so the naming is provisional and temporary—yet no less powerful for all that. In the Prāsaṅgika Madhyamaka school of Buddhist philosophy, often considered the pinnacle of Buddhist understanding of the phenomenal world, there is no essence, ground, or basis for any object's existence outside the act of naming or labeling it. Naming brings the object into being as an object of consciousness, and is immensely powerful for this reason.

The Viral Power of the Fragment

But not all montage aims at producing a new form. Jean-Luc Nancy recently commented that DJ mixing could be seen as "putting together two forms that start out as heterogeneous or foreign to each other, something like collage. This attests more to the degree of insecurity or instability of the forms in question." On the other hand, mixing can also be viewed as "constitution of a new form."[12] There is no inherent need for a montage to be "finished"—or, for that matter, "framed." In fact, part of the power of montage relates to the peculiar nature of fragments as vehicles of contagious mimetic energy, and the possibility that one can play with fragments in such a way that the active viral power of the fragment is not limited by being too quickly absorbed into a new fixed form. Even in cinema, where the search for a definitive "director's cut" can evoke a kind of Holy Grail, there are instances of a productive hesitation in fixing a particular organization of cuts and sequences—for example, Orson Welles's "unfinished" *Mr. Arkadin,* which exists in five or more versions, all of which contain sequences not found in the others, all of which have been described as "butchered"; or the later career of New York underground filmmaker Jack Smith, who would edit and reorganize his films while he screened them, so that his collective oeuvre consists of an ever-shifting rhizomatic mass of clips and cuts taped together on the fly.

By breaking open a named, coded form, one creates fragments that are unfinished, charged with some of the energy of the form they came from, whose likeness they still contain. Think of hip-hop samples, many of which are barely more than a note or a beat, but which still signify a trace of the pattern that they were associated with. A virus integrates into a host cell in order to replicate; it is unfinished and open, since it has no reproductive powers of its own. A fragment is an unstable unit—but we, too, are "unstable units"; and our longing for wholeness, our need to populate our equally unstable environment with wholes, expresses our discomfort with that moving, shifting chaos that I described in the chapter on transformation.

Robert Farris Thompson, a theorist of African art, notes the importance of breaks in various African art forms: "Just as a break in the surface of a rock emphasizes the emergence of a painted spirit in South Africa, and a break *(kasé)* in drumming occasions possession by the spirit in Haitian vodun, so a break in jazz drumming often inspires virtuosic leaps of imagination in the playing of a soloist."[13] In hip-hop, the "break" is that fragment of an old record where the sound strips down to drums, a fragment that can be looped to build up new rhythm tracks of potentially infinite length. But the word also contains the meaning of a rupture from which a dizzying energy emerges, as in the "breaking" of breakdancers.

The destructive part of montage, the part connected to destroying an image which holds together a form of power and naturalizes it, often involves defacement. Many montages take the form of defacements: think of the way *Adbusters* magazine defaces corporate ads with its "subvertisements," or satirical images of George W. Bush, or John Heartfield's cut-ups of Nazi propaganda. This kind of montage damages, abuses images of power by interrupting them with an image or a word from elsewhere. The cover of the Sex Pistols' *God Save the Queen,* by Jamie Reid, is a perfect example of this. It shows Queen Elizabeth's face, and the eyes and mouth of the image have been covered over with a montage of letters spelling out the name of the group and the title of the record. Literally, a de-facement.

In his book on the topic, Taussig observes that "defacement is often the first thing people think of when they think of mimetic magic, like sticking a needle in the heart of a figurine so as to kill the person thereby represented, and it is no accident that this was Frazer's first example in the scores of pages he dedicated to the magic art in *The Golden Bough.* Defacement is privileged among these arts of magic because it offers the fast track to the mimetic component of sympathetic magic." To spell it out: violence is the fast track because "it's not only as if disfiguring the copy acts on what it is a copy of, but that, associated with this, the defaced copy emits a charge which seems—how else can we say this?—to enter the body of the observer and to extend to physically fill, overflow, and therewith create an effusing of proliferating defacements."[14] This defacement is contagious, as we see in the excellent graffiti documentary *Style Wars,* where graffiti rapidly spreads in 1970s New York from the block-celebrating marker pen of the originator, Taki 183, to massive spray-paint works that cover entire subway trains. The contagion also spreads to those who are against graffiti, such as the bully named Cap, who is committed to covering every "burner" with his own signature and thus defacing the defacement of the graffiti, just as the Sex Pistols were beaten up for defacing the queen and ridiculing her Jubilee.

The breaking up or breaking open of a form, as in many of the above examples, is said to liberate energy, which is "contagious," "viral," "infecting" other forms. What exactly is this energy, and how does it relate to copying? A comparative study of historical and cultural figurations of energy remains to be written. It would help to illuminate many of the questions concerning copying that I have set out, since I have frequently had to refer to "mimetic energy" without being able to clarify exactly what I mean. Such a study would have to examine various models of energy in physics, and compare them to Chinese *qi,* Tibetan *rLung,* and Indian *prana.* Without claiming that these words all refer to the same thing, one can say that mimesis plays a crucial role in all such theories, since the search for the fundamental building blocks of reality is almost inevitably a mimetic one. This applies equally to traditional or religious models, to the reductionisms of contemporary neuroscience, and to the laws of physics and the quest for a theory of everything. These words serve to track or label qualities of flux, transfer, transformation, impermanence in matter, and they signify a certain inherent openness of matter which can be worked with through the practices of montage. Thus, the label "energy" is linked to an immanent and pervasive nonduality, to nonsensuous similarity, and to the maternal sameness from which all name and form emerges and into which it disappears.

Humor

When I teach montage, students tend to bring up examples of political humor that use montage techniques, such as clips from Jon Stewart's *Daily Show* showing Bush as governor of Texas in debate with Bush as president. Montage makes us laugh—think of *Monty Python's Flying Circus,* with its Dada-like animations of Victoriana, British Empire memorabilia, and psychedelia. In trying to understand why montage should be funny, I am reminded of Bergson's famous dictum that "the attitudes, gestures, and movements of the human body are laughable in exact proportion as that body reminds us of a mere machine."[15] Laughter reveals a disjuncture between the suppleness of human cognition and an outward appearance whose rigidity or clumsiness does not accord with the expectations of that cognizing consciousness. We laugh when something we assumed to be a rock-solid state of being is revealed to be other than what we thought it was. Something slips. Taussig expresses amazement at "how naturally we entify."[16] We laugh when we slip from one entification to another, when we see the slip happening—in watching someone slipping on a banana peel, in the defacement of someone being hit with a custard pie, in a montage in which one gesture is superimposed on a gesture of a different kind (such as Bush the governor speaking to Bush the president), in a slip of the tongue that contains a sexual meaning. Laughter is a recognition of the truth manifested in that slip, which is the slip that undoes the apparent permanence of the things of this world; and thus, as Simon Critchley notes, it is "an acknowledgement of finitude."[17] Puns are funny because they reveal, at the level of the unit of semantic meaning "itself," the possibility of radical disjunctures and breaks. They show that a word is an unstable montage of meanings, held provisionally in place by a particular dominant meaning, but with a Deleuzian swarm of virtualities always ready to resolve at the level of the actual, to become present. Finally, "I" laugh—and "my" laughter is the recognition and reflection of the instability and impermanence of that "I," the "I" that is also a "mere copy," "my" greatest and most obsessive fabrication.

Montage as a Way of World-Making

Historically, montage and collage as modernist aesthetic practices date to the early decades of the twentieth century. The scientific context for the "discovery" of these methods includes the formalization of a number of scientific systems which consist of the permutation, by chance or otherwise, of various basic components: chemistry's periodic table of the elements (Dmitri Mendeleev, 1869), atomic physics (the electron, the nucleus, quantum physics, the isotope, 1897–1918), genetics (Gregor Mendel, 1866), and set theory (Georg Cantor, 1874). Technologically, the advent of the camera (1826), the telephone (1876), the phonograph (1877), and the movie camera (1889) are also highly relevant. To this we could add a number of historical and political developments—Walter Benjamin, for example, points to aerial warfare in World War I as having a key effect on the destabilization of linear narrative.

Yet histories of montage still limit themselves to a very particular and at this point predictable framing within the arts. From Picasso's and Braque's first *papiers collés,* montage questioned the prevailing categories of art and life—by directly introducing materials from everyday life (newspaper items, found objects) into the painting, sculpture, or poem. The trajectory of the arts in the twentieth century, from traditional forms to the dematerialization of the art object in the happening, the installation, and the intervention, has occurred via a working-through of montage.

When did it first become apparent within the tradition of modern art that montage was not only a particular aesthetic technique, but a way of understanding the world? In 1958 artist Brion Gysin and writer William S. Burroughs developed the concept of the cut-up. Gysin discovered the technique when he slashed through a pile of newspapers in his studio with a Stanley knife and started to read across the slashed pages. Burroughs applied these techniques to his own writing, using his own work and "found" material from Shakespeare and Conrad, finding in the technique a method of generating new meanings that tore the original texts free from the control structures of conventional grammar and semantics. Gysin and Burroughs also applied the technique to film and sound (though there is a rich tradition of montage in older media). But Burroughs' real insight was that reality itself could be a viewed as a film, a set of recordings, or a montage, because—posited in Platonic terms of unchanging essences, or in Burroughs' terms as endlessly replayed recordings—reality is a fabrication made by certain parties who have an interest in presenting this fabrication as "natural." "'Reality' is simply a more or less constant scanning pattern—The scanning pattern we accept as 'reality' has been imposed by the controlling power on this planet, a power primarily oriented towards total control."[18]

But this fabrication, which has been naturalized and presented to us as "common sense," can likewise be cut up, transformed, and revealed to be a fabrication. As Alain Badiou notes, ideology itself is a montage, a fabrication or fiction, and it can be torn up, destroyed, rearranged, and refabricated.[19] Timothy Murphy points out that Burroughs and Gysin's theories have much in common with those of the French Situationists, who in the mid-1950s in Paris also proposed strategies of montage and appropriation of mass cultural elements *(détournement),* aiming to take apart an imposed spectacle that presented itself as reality. For the Situationist Guy Debord, "the spectacle is *capital* to such a degree of accumulation that it becomes an image"; thus, it is reality as it appears in modern society that is the montage, and Debord proposed an "ultra-détournement" based on the observation that, "ultimately, any sign or word is susceptible to being converted into something else, even into its opposite." Therefore, "when we have got to the stage of constructing situations—the ultimate goal of all our activity—everyone will be free to detourn entire situations by deliberately changing this or that determinant condition of them."[20]

The earlier political-montage artists of German Dada and the Soviet avant-garde saw montage as a way of attacking the ideologies and worldviews of the ruling classes, but it is doubtful that they thought montage itself constituted the mechanism of both control and liberation, in the way that Burroughs and Gysin or the Situationists did. Vertov, for example, writes that "Kino-eye uses every possible means in montage, comparing and linking all points of the universe in any temporal order, breaking, when necessary, all the laws and conventions of film construction"; but this is still a statement concerning the possibility of cinema, rather than a generalized observation about the nature of perception.[21] A combination of psychoactive drugs, phenomenology, and psychological and spiritual introspection allowed the artists and activists of the 1950s to recognize the way that cognitive processes in general resemble montages.

Conversely, we might say that today montage is a key practice or strategy of late capitalism. It is a way of implementing industrial techniques for gathering and processing masses of things as copies of one another through a system of choices and decisions which allow for permutations within a predefined network. This applies as much to democratic party politics, which offers an endless variety of candidates who are "all the same," as it does to Starbucks' endless varieties of coffees, sizes, accessories. For Baudrillard and for the Situationists, this leads to a duality in the concept of montage and of copying: the simulacrum and the spectacle on the one side, seduction and *détournement* on the other.

Montage in "Classical" Non-Western Cultures

Equally important, Burroughs and Gysin's claims for montage reveal that a great range of traditional practices, though not usually understood in such terms, are deeply related to montage. Gysin linked his own discovery of the cut-up to traditional Moroccan magic—to the assemblage of words and various kinds of matter that one finds in a spell.[22] Contrary to most histories of montage or collage, one can find elements of montage in medieval Christianity, as well as in Buddhism, Taoism, and shamanism, and doubtless in other traditional cultural forms too. Picasso's discovery of collage has been linked to his exposure to African art; but what this really means is that Picasso appropriated existing traditional practices of montage and collage from the complex religious, aesthetic, and political situations in which they occurred.[23] In the general history of montage, not to mention the history of copying, modernism occupies but a single chapter, fairly late in the book.

Indeed, although most of the examples I have given so far have come from modernist Western art, one could argue that what is called montage is the prevalent form that art takes in any culture not built around the assertion of essence. In *Ten Thousand Things: Module and Mass Production in Chinese Art,* sinologist Lothar Ledderose catalogues the long history of the use of modules in Chinese art, dating back to the techniques used for fabricating funerary bronzes around 1000 B.C.[24] Ledderose shows the variety of ways in which the Chinese have used devices—such as molds, movable type blocks, and stencils—to produce multiple, standardized elements which can be combined with other such elements to fashion complex, varied artifacts. Molds were used to make bronzes and to fabricate the famous terracotta army; type blocks, to produce scrolls and books; stencils, to make religious art. These methods produce artifacts in sets, which are copies of each other, but copies that are never identical, due to variation, both in the combination of standardized elements selected for any particular artifact, and in the specific configuration or placing of these elements. Conversely, the *I Ching,* the great Chinese divining system, consists of sixty-four modules, each describing a particular configuration of natural forces. These sixty-four modules contain the totality of the possible states in which nature can configure itself, as well as the possible movements between modules, implying a modular approach to phenomenality.

Modularity in this sense can be observed in a variety of cultural forms around the world. Specific ragas in Indian classical music, for example, are built around a combination of a particular scale, rules for moving around the scale, and motifs or ornaments that are used in particular ways. In terms of Western musicology, a raga is confusing, since it cannot be defined as a particular composition (in fact, there are thousands of compositions that can be made within any particular raga), yet it is more than just a scale in which one can freely improvise (there are an almost infinite number of ways of improvising within the raga, but there are also rules which structure the improvisation, prescribing allowed movement between notes, and the type of rhythm, tempo, and mood). The polyrhythms that are found in a variety of African musical forms are also modular in this sense: "The most important formal element in African music is that instead of having a single meter . . . a performance puts two or more different meters together, as if one drummer were playing in waltz time and another in march time, for example. Rhythm is also based on contrasting recurrent beats with irregular patterns."[25]

Food as Montage

The cultures of alimentation offer another example of montage operating at a fundamental level of human activity. In *Empire of Signs,* his strange semiotic meditation on a trip to Japan, Roland Barthes observed that the bento box or sushi plate turns the act of eating into a montage, whose order you construct in the moment of selecting this or that from the plate in front of you:

> Eating remains stamped with a kind of work or play which bears less on the transformation of the primary substance . . . than on the shifting and somehow inspired assemblage of elements whose order of selection is fixed by no protocol (you can alternate a sip of soup, a mouthful of rice, a pinch of vegetables): the entire

> *praxis* of alimentation being in the composition, by composing your choices, you yourself make what it is you eat; the dish is no longer a reified product, whose preparation is, among us, modestly distanced in time and in space.[26]

But there are many food regimes that relate to practices of copying, and more specifically to the making and consuming of food as an art of assemblage. Recipes are themselves something like manuals of montage, advising how to cut an object, subtract from it or add to it, how to combine it with other things, how to transform the ingredients through heating, how far the transformation should go, the ratio of "raw" to "cooked" of different ingredients, and how to arrange things on a plate and serve them. But everyone knows that far from being a reification, a recipe will itself undergo processes of experiment and transformation which will also be shared with other people. And curiously, in recognition of this, U.S. law does not permit recipes to be copyrighted, unless they involve "substantial literary expression."[27]

Is there any way that food could *not* be a cut-up? Perhaps in the case of an anaconda who swallows a dead sheep whole and then slowly digests it, but then we are in a world that we go to great lengths to separate from "ours" (there are often different vocabularies for human and animal alimentation: for example, in German, *essen* versus *fressen*—"to eat" versus "to feed," as at a trough). The inventory of the processes of preparation, assimilation, digestion, and incorporation—selecting, cutting, combining, cooking, eating—are very similar to the various processes of montage I have set out above. Cooking techniques imitate the basic processes of assimilation of food: ingestion through the mouth, the cutting and grinding actions of the teeth, the work of digestion in the stomach and the GI tract. The imitation of these corporeal processes through culinary technologies allows the taboo on transformative mimesis (described in

Chapter 3) to be implemented and managed, copying the disturbing plasticity of this fundamental corporeal mimesis in an attempt to fix it as cultural and technological, as some *thing*. Our bodies are the product of these varied processes of assimilation. The object produced by this montage is: us.

Montage as a Feminist Practice

In *Cunt-Ups*, her pornographic parody/pastiche/appropriation of Burroughs and Gysin's cut-up method, Dodie Bellamy asks: "Is the cut-up a male form? I've always considered it so—needing the violence of a pair of scissors in order to reach nonlinearity."[28] But many of the greatest montage artists—Hannah Höch, Barbara Kruger, Kathy Acker, Delia Darbyshire—are women. And many stereotypically feminine arts—cooking, quilting, knitting, sewing—are profound developments of the montage principle. If these arts are derided, considered second-rate or derivative, mere crafts, isn't this because they are associated with copying? Because they "merely" rearrange preexisting materials, and are thus seen as lacking the originality and authenticity of heroic fine arts such as painting and sculpture? The cut-up, then, according to Bellamy, is a matter of appropriating this "feminine" principle and turning it into a method, the mimicry of montage, the Platonic representation of the non-Platonic plasticity of mimetic energies and forces. In this view, violence is an attempt to imitate and thus control the transformative flux that is always already there, and that men are so afraid of. In Eisenstein's early writings on montage, for example, one senses a kind of terror at the chaotic openness that is immanent in montage, and that Eisenstein tries to contain by his insistence on the propagandistic political organization of montage fragments, and the strange idea that cinema is a "fist."[29]

But if one does not "need a pair of scissors in order to reach nonlinearity," this suggests that the realm of the feminine (however we wish to frame it) is already a realm of montage. Pursuing this line of thought, we might reach a place where montage itself is cut into so many pieces that it no longer makes sense to call it montage.

Consider sexual reproduction as montage. The proximity of bodies and minds; insemination; the cutting-up and combining of the mother's and father's DNA. Then the slow process of transformation and growth of the fetus, connected to the mother via the placenta, through which nutrients and waste products are exchanged and filtered. Differentiated but undifferentiated; mother and unborn child; both one and not one. The kicking of the baby inside the mother. Uterine contractions, labor, and the birth of the child, followed by the cutting of the umbilical cord. The partial object of the mother's breast during the months after birth. How much of this process should we recognize as montage?

If we do not call it montage, this is because the cuts happen too slowly; or invisibly; or with too great a degree of complexity, in too many small, almost simultaneous ways. Montage, in the cinema and elsewhere, is a matter of tempo. Violence, for Bellamy, means a crude, blunt, too-rapid forcing of things. Gayatri Spivak, in glossing Jacques Derrida's term "teleopoiesis," likewise insists on a practice of "copying (rather than cutting) and pasting."[30] Do we have to speak of cuts in order to understand montage? It has been argued that throughout Dada and Surrealism, from Tristan Tzara through André Breton and Max Ernst, there existed an erotics of montage built around the intimacy and proximity of organically unrelated objects and entities, as well as an erotics of collaboration.[31] Luce Irigaray speaks of a "placental economy" that mediates self and other.[32] The placenta modulates a boundary between mother and fetus, allowing nutrients and hormones to move back and forth, also acting as a barrier permitting the coexistence of self and other. But the umbilical cord is literally "cut" at birth, and the placenta is expelled from the wall of the mother's uterus. The model of autopoiesis proposed by

Francisco Varela and Humberto Maturana, emerging out of a context of biology, systems theory, and Tibetan Buddhism, provides a way of understanding how a recognizably self-similar entity that is nonetheless in a state of constant flux maintains a relatively stable form through a continuous process of exchange with its environment: the organism as biological montage.[33]

The Limits of Montage?

Theodor Adorno delivered the following damning critique of montage-based art: "But montage disposes over the elements that make up the reality of an unchallenged common sense, either to transform their intention or, at best, to awaken their latent language. It is powerless, however, insofar as it is unable to explode the individual elements. It is precisely montage that is to be criticized for possessing the remains of a complaisant irrationalism, for adaptation to material that is delivered ready-made from outside the work."[34]

When I said that "various elements" are montaged together to make up reality as it appears to consciousness, the question does arise as to what these elements are. As anyone who has looked at a lot of modernist montage or permutation-based art knows, there are limits to the power of an art that can only dispose elements, be they images, texts, etc. Adorno's criticism is valid for certain iconographic/clastic modernist art forms, and accounts for some of the contempt that many of my students have for the political montages in *Adbusters.* It is also a valid critique of the way that montage has been integrated into the logic of late capitalism: as a series of multiple-choice questions that simulate a process of consultation in which all possible or allowable answers to questions have been generated in advance.

But does Adorno's critique apply to all art or all uses of montage, particularly those developed by folk cultures? I don't think so. The nonrepresentational patterns that are cut up and mixed together for quilts, or the beats and drum sounds that are such important elements for a DJ mix, are often composed of elements that are so abstract, basic, and monad-like that it is meaningless to speak of them as material, or as having a "latent language." These units, used in large numbers, and in ways that link them to time, to contingency in all its forms, come to possess emergent properties. In the same way, language itself could be regarded as a montage of elements called "letters," and life as a montage made of biological and physical elements. It is clear that complexity can emerge from the permutations and combinations of such basic elements, and part of that complexity is the result of the provisional nature of any attempt at defining "elements."

What, after all, are "elements"? This is a topic I have already touched on briefly in Chapter 2, in discussing copia as multiplicity, and in Chapter 3, in discussing the units of universal repetition, including the atoms of Lucretius and Dharmakirti. Claude Lévi-Strauss speaks of the *bricoleur* as follows: "The rules of his game are always to make do with 'whatever is at hand,' that is to say, with a set of tools and materials which is always finite and is also heterogenous."[35] Lévi-Strauss claims that the ensembles of such materials are restricted to a certain set of permutations and juxtapositions prescribed by the history of each object or, more broadly, by tradition—as we have seen in quilting, cooking, and various folk arts. But Lévi-Strauss is unclear about these elements. Isn't his assumption that these elements are permanent and unchangeable, key to formulating folk thought as "myth," in the sense of superstition to be explained as "structure"? And wasn't it on this particular point that Derrida made his critique of Lévi-Strauss, more or less demolishing the intellectual basis of structuralism?[36]

One could make a similar critique of Burroughs or Debord—or Baudrillard's *Simulations*, or the film *The Matrix*. Each of them posits a secret structure underlying the appearances that constitute everyday life. All vacillate between claiming that this structure is an imposition of a particular political power, and saying that this structure is reality itself. But their claims that reality is a montage do not go far enough, since, as Adorno says, they retain a faith that underneath this particular figuration of reality there are solid elements, and a real, substantial reality.

The problem remains unresolved today. As we have noted, montage is key to the impasse of what is called "postmodernism" or "late capitalism" or, today, "globalization."[37] The Derridean trace and the Deleuzian assemblage were part of an anti-essentialist emancipatory project that, as thinkers like Badiou, Žižek, and Peter Hallward have recently pointed out, has instead been absorbed into capitalism. But Badiou's reworking of set theory as a basis for thinking "pure multiplicity," and Žižek's "parallax view," are themselves ways of rethinking montage, particularly with regard to what limits an assemblage, and how truth can emerge from the experimental production of new multiplicities. Moreover, just as in Buddhist philosophy the relation between the absolute view of nonduality and the relative view of conventional name and form remains a matter of considerable dispute, Badiou has so far been unable to persuasively theorize the gap between the pure multiples of Being and the event, despite his affirmation of what he calls truth as "process" or "passage" (which is perhaps a way of saying "montage") between them. The question remains: Do we believe, as Burroughs and Gysin postulated, that "nothing is true, everything is permitted," in an infinitely permutating series of assemblages? Or is there a limit? And if so, what is it?

The great Balinese gamelan master Wayan Lotring once said, "In my time, all music was nothing but nuances."[38] By "nuance," he meant the recognizable mastery of a form, along with the ability to reveal the multiple variations of which that form is actually structured. And to reveal these variations not merely as mathematical permutations, but as affect or emotion-laden structurings of consciousness and experience that are highly specific and at the same time transient and impermanent—"copies" if you like, but contingent, evental copies that emerge from and fade back into the emptiness which is their ultimate nature. Adorno speaks of the need to "explode the individual elements"—but the very violence implied in this statement implies that Adorno still believes in the Platonic reality of such elements and remains trapped within a historically contingent cognitive prison. "Elements" themselves are interdependent constructions or fabrications that reciprocally define themselves in the act of being juxtaposed—and this is equally true of the twelve-tone scale of the serialists, or the positing of all possible sounds as music by post-Cagean composers such as Philip Corner. Elements emerge, situationally, in the moment of their being perceived in a particular environment. As human beings inhabiting a relative world, we are constantly involved in the act of fabricating, configuring, and sustaining such elements. Insofar as anything "is," it is a montage. But then, how could montage be an intervention, since what it ostensibly interrupts is also a montage?

Digital and Analog

The question as to the nature of the elements that make up a system of combinations is key to understanding computers, whether digital or analog ("digital" connotes the abacus; "analog," the slide rule). The digital computer is of course a mimetic machine par excellence, allowing the representation of anything that can be coded as a series of ones and zeros, and of whatever can be made through the manipulation, combination, permutation of that code and what it represents. "Cut and paste," one of the key metaphors found on a personal computer, refers to an operation of montage or collage, even though there is no literal cutting or pasting on a computer. Historically, the philosophical and material innovations that led to the development of the digital computer parallel those in montage. Such innovations include the generation of set theory as a formal language for describing objects, procedures, etc. by Georg Cantor and others at the end of the nineteenth century; the work of Wittgenstein's colleague Alan Turing, between the wars, to develop a machine capable of representing everything in the universe; and the extrapolation of this work after World War II by cybernetics and systems theorists, so that by 1954 Norbert Wiener could write: "A pattern is essentially an arrangement. It is characterized by the order of the elements of which it is made, rather than by the intrinsic nature of these elements. Two patterns are identical if their relational structure can be put into a one-to-one correspondence. . . . A copy of a painting, if it is accurately made, will have the same pattern as the original while a less perfect copy will have a pattern which is in some sense similar to that of the original."[39] The overall trajectory goes from an intense but highly specific practice of the manipulation of combinations, to a generalization of this practice that includes almost all domains of human activity.

What is the specificity of digital "cut and paste"? Analog machines use a physical quantity to represent something, whether a sound, a word, a number, or anything else. Digital machines use numbers—ones and zeros. When we speak of "digital copying," we are usually thinking of digital files made of scanned or sampled objects—for example, MP3 audio files, JPEG images, or .mov film clips. The word "sample" comes from the way in which an analog impression of a sound or other source is made and then converted to digital data. While a photograph involves exposing a piece of plastic film coated with light-sensitive silver halide salts for a controlled period of time to light rays reflecting off an object or scene, a digital camera exposes the same scene to a grid of light-sensitive photo-transistors which are translated into a stream of data that takes the form of ones and zeros. While a magnetic tape or vinyl disc records a continual real-time flow of sound onto a recording surface, creating an analogous (i.e., analog) sound, a digital sound is sampled thousands of times per second, meaning that a "snapshot" is taken many thousands of times per second of the sound source, which is converted into digital code. Each snapshot is then sequenced and played back, in sequence, very fast (a CD has a sampling rate of 44,000 hertz, meaning that 44,000 snapshots are played back, in sequence, per second). We sample a sound by making a number of copies of it and playing them back in time. We sample an image by taking a number of very tiny snapshots of it and playing them back, side by side, on a computer screen. Thus, all sampled objects are, in effect, montages and partake of the same viral power that montage has—which is the power of the fragment, the unfinished, discontinuous partial object.

This discontinuity is exploited to powerful effect on peer-to-peer networks such as BitTorrent, where the copying of a digital file lacks even the narrative of a one-to-one copying (which is always to some degree a myth), and where the object is reconstituted through the assembly of a multitude of almost-simultaneously produced fragments. A digital file is both a pattern and instructions for making a pattern.

In a 2002 essay entitled "On the Superiority of the Analog," Brian Massumi defines "analog" as "a continuously variable impulse or momentum that can cross from one qualitatively different medium into another. Like electricity into sound waves. Or heat into pain. Or light waves into vision. Or vision into imagination. Or noise in the ear into music in the heart. . . . Variable continuity across the qualitatively different: continuity of transformation."[40] In other words, "analog" describes all kinds of mimetic transformations, regardless of whether they involve machine technologies such as the cassette recorder or the phonograph, which are traditionally associated with the word. It is the ultimate nonduality of these states of transformation that makes them possible at the relative level—thus the slightly awkward semi-conceptual terms "continuously variable impulse" and "momentum" that Massumi is forced to use in describing what, finally, is beyond description.

Massumi defines "digitization" as "a numeric way of arraying alternative states so that they can be sequenced into a set of alternative routines" (137). Massumi points out that digital technologies manifest their effects only through the analog; thus, even though a word processor turns letters and words into code, we can access the code only through its translation to analog via a computer screen. The same is true for "digital sound": "It is only the coding of the sound that is digital. The digital is sandwiched between an analog disappearance into code at the recording and an analog appearance out of code at the listening end" (138). In fact, even this minimization of the digital is too generous, for all digital code exists solely materially, in analog form—for example, in the form of a series of electrical charges in storage media such as a hard drive. "Digital" describes a particular set of very complex analog routines that allow for the generation of potent symbolic languages; but there is no digital realm outside the material-yet-groundless mimetic traces and processes by which what we call "code" is disseminated.

Is there, then, a fundamental difference between analog media such as the photographic negative, magnetic tape, or vinyl record, which retain some form of direct impression of the light or sound waves that they are touched by, and digital media, which have to convert the same moment of touch into digital code, for subsequent storage and retranslation back into a visual image or sound? Many people aside from Massumi have argued in favor of analog media as presenting a more vibrant, alive-sounding copy of a sound; Thurston Moore, for example, speaks of there being an "analog heart." The cassette head that "reads" tape actually touches the tape, thus forming a tactile mimetic link with the copy, while a hard drive read/write head flies over the hard disk without actually touching it. But recent informal studies by musicologist Jonathan Berger suggest that those familiar with compressed digital sound files actually prefer them.[41] And there is no philosophical basis or ground on which to fundamentally separate digital and analog sound-recording technologies: they are significantly different, but they both involve disseminative, mimetic processes that are material as well as interpretive.

How does digital montage fare in relation to Adorno's critique? At first, one might think that the "elements" of digital montage are code (binary or otherwise), and code is ideology insofar as it structures all possible configurations and iterations of those elements. But since all digital code acts as a label or marker for the analog, the complexity of the elements that compose the montage is not necessarily lost (think of the relation of digital to the complex matrix of analog bodies, objects on Craigslist, for example), and the possibilities for the patterning of provisional, material elements increase exponentially. While it is possible to manipulate analog media, by painting onto a negative, or splicing a tape, one is also limited by the materiality of analog media, which set certain limits on how one can access the data. Once an image or sound is stored as digital code, it can be manipulated in a much larger number of ways. The montage cut-and-paste effects that can be achieved with scissors or razor and a print image or magnetic tape are rather crude compared with similar effects obtained with digital software, because an image can be edited and rebuilt from the ones and zeros. Thus, the world of Photoshop, Cubase, Final Cut Pro, and other sample-processing softwares is a world in which the power of montage, the creator of discontinuous mimetic effects, is resplendent. The digital may be necessarily "sandwiched" between analog disappearance and reappearance, but the sandwiching remains decisive as a way of opening up possibilities for copia. Industrial folk cultures quickly recognized this potential of computers and appropriated it (for example, the use of digital samplers in hip-hop or hardcore, drum-and-bass, and their many variants), practicing montage in awareness of the provisional, apparitional nature of all elements and materials involved. No doubt there are many reasons (and also none) why hip-hop producer J Dilla gave the title "Donuts" to the last record and sampladelic masterpiece he issued before his untimely death; but this word captures perfectly the sweet and tasty, holey emptiness of the digital sample manipulated with great subtlety in time (which, according to a song title from the record, is to be understood as "the donuts of the heart").

Modern, Postmodern, and Amodern Montage

The framing of digital-code manipulation as a "cut-and-paste" act of aesthetic montage expresses perfectly the way an aesthetic façade is used to ideologically justify the disjunctures of capitalist modalities of production, including the production of subjectivity. Thus, what began as a critique of the capitalist production of such disjunctures and the trauma associated with them has today become their justification, even their modality.

But as we have seen, it is not accurate to define "montage" only as an art form, and in particular one associated almost exclusively with a certain high-modern history of art that begins with the collages of Picasso and Braque in France in 1912, or the photomontages of Höch and Heartfield in Berlin in 1919, or Eisenstein's cine-montage, or the Surrealists' literary montages, or Pierre Schaffer's discovery of *musique concrète* in 1948. Historically and otherwise, montage and collage are not fundamentally aesthetic practices; they are part of the mimetic excess that is found throughout every society. Framing them aesthetically limits their radical potential and the potential for a transformation of the societies we live in. Even the phrase "cultures of copying" brackets off the radical transformative potential of mimetic play as belonging to a certain order of leisure, entertainment, and private enjoyment that is part of the dominant ideology today. Nevertheless, the popular practice of montage continues.

After the destruction of the World Trade Center towers in Lower Manhattan on September 11, 2001, one of the most surprising developments was the appearance of spontaneous memorial shrines around New York City. They seemed to grow out of the proliferation of art and writing, much of it photocopied onto 8½-by-11 paper, which covered the bus stops, the closed-up storefronts, and the walls of available public spaces in Lower Manhattan in an enormous collectively produced montage of poetry, drawings, inscriptions, and missing-persons photos of the most varied kind. The shrines, which appeared in public parks such as Union Square, consisted of spontaneous arrangements of flowers, pieces of cloth, candles, religious icons, flags, and other American icons, and a vast montage of graffiti and inscriptions, spread out on paper, cardboard, and other surfaces. The shrines often grew out of the ground in the form of an ascending spire or an altar, but they also accumulated dirt, decaying organic matter, melting candle wax—they exhibited the polarities of the sacred.

I do not mean to be facetious when I propose that the event of 9/11 can be understood as a vast unleashing of mimetic forces. Even at the level of spectacle, many people were struck by the compulsive repetition, in the broadcast media, of a video clip showing a plane crashing into one of the twin towers—a repetition that continued for months until criticism by viewers brought it to an end. The incessant video broadcasts were a perfect illustration, in the global media, of a tendency first described by Freud: the compulsion to repeat in the face of trauma. Indeed, the "twin towers" were themselves mimetic, their repetition a signifier of the vastness of the economic and political forces that they embodied and facilitated on behalf of "world trade." The repetition of their destruction was also doubly traumatic because it blocked a certain psychic release that would have been possible if people had been able to call the destruction a singularity, a fluke, or a "lucky break" (for the terrorists), as one of my friends described it. The explosive juxtaposition of airplane and building was a colossal act of defacement, of montage, if you will—and perhaps this is why Karlheinz Stockhausen, one of the most venerable practitioners of musical montage, bizarrely claimed that 9/11 was a great work of art. And of course 9/11 provided a kind of mimetic reservoir that could be called on during the fabrication of the war on Iraq and the War on Terror, both major episodes in the history of mimesis as deceptive power.

Fast forward a few years to January 2007. The scene: the Tate Britain museum in London. The venerable late-Victorian exhibition hall is displaying a very similar arrangement of materials: photocopies of faces, slogans, written texts, assembled in a chaotic cluster that fills the room. The piece, entitled *State Britain,* by British artist Mark Wallinger, is a re-presentation of a 2002 antiwar protest that protesters, led by Brian Haw, installed on the grass of a traffic island directly opposite the Houses of Parliament in response to Britain's participation in the Iraq War. The protest, gritty, chaotic, and disheveled, defaced the polished historical legitimacy of the Parliament buildings. An eyesore intended to evoke the deceptions of the British government's participation in the war for American empire, it led to the passing of a law banning all unlicensed protest within a one-kilometer radius of Parliament Square, and to the subsequent dismantling and evacuation of the protest. In accurately recreating the protest, Wallinger repositions it as an art object in the long history of modernist political montage that began with German Dada and Soviet cine-montage. Fortuitously, the installation itself rests on the very perimeter of the area in which protest has been proscribed, and this perimeter is marked in red tape, literally splitting the installation into segments of legitimate and illegitimate enunciation.

As much as I admire Wallinger's work, one cannot claim, as nearly every history of montage or collage does in similar situations, that here, once again, the artist has perceived the aesthetic or even political value of what was at best an example merely of folk art or everyday practice. Haw's protest is already an aesthetic-political artifact which manifests the formal qualities of montage. But the framing of montage as an aesthetic practice is precisely what renders it a part of bourgeois ideology, and what blocks the more powerful liberating energies that are immanent to montage, no matter what temporary provisional or strategic gains may be made from work such as Wallinger's. And it is here that the 9/11 memorial shrines—which also brought together in assemblages a variety of materials and practices, without its being necessary or even possible to reduce them to an aesthetic (or for that matter political!) framework—are the herald of a very different vision of montage. It is, to use Latour's term, an "amodern" vision, meaning neither modern nor premodern, but inhabiting a continuum that encompasses both and more. Like Haw's demonstration, the 9/11 shrines appeared outside the gallery or museum, without the alibi of art. Unlike Haw's demonstration, they existed as temporary, anonymous, collective assemblages, and they were found in out-of-the-way corners of the city, next to trash cans and doorways, as well as in larger public spaces such as parks. They consisted of performance, music, social hanging out, and discussion as much as material artifacts—the return of Allan Kaprow's "happening" as a montage of events, this time without the frame of art. If anything, the shrines were closest to religious festivals or places of pilgrimage, where people likewise leave graffiti and other material traces, amid the same kind of decidedly unorthodox religious heterogeneity. But this, too, brackets off what happened. To see the world, the self, and the community as montage, and to live the consequences of that vision with openness, without assigning it to a particular domain of human activity—this is the freedom toward which we are heading.

6/The Mass Production of Copies

President Carter loves repetition
Chairman Mao he dug repetition.

—The Fall, "Repetition," *Bingo Master's Breakout,* EP disc (1977)

The Multiple

You can see them on the factory conveyor belt in Charlie Chaplin's *Modern Times.* What are they? Rectangular slabs of metal with two metal knobs growing out of them. They emerge out of a machine in an apparently infinite number, all exactly the same. They have no identity or purpose other than to make life hellish for the factory workers whose job it is to assemble something from them, but they are clearly copies. The workers themselves become twitching machines, each devoted to a single action on the assembly line, a gesture they repeat endlessly until it is all they can do in the world.

So far, we have examined different ways of framing a single act of copying, or a single event in which a copy manifests. But as the origins of the word "copy" (in "copia," "copiousness") attest, implicit in the notion of copying is the possibility of producing multiple copies. If nothing else, we all understand that "copy" today means "more than one." Copying is an act of repetition, and contains in it the possibility of repeating that repetition unto infinity. If the world we live in today is obsessed with copying and copies, it is because that world is one which is based on the amazing realization that we (who are "more than one") can make "more than one" of just about everything, and, more darkly, that we are interested only in things that we can make, buy, or sell "more than one" of.

If marketplaces have always been places which heap up piles of objects in a display of richness, today's supermarkets and malls, with their endless lanes and canyons stacked with vast quantities of goods are like galleries or cathedrals of copying. We walk these lanes in a strange trance, hypnotized by the sheer numbers of identical objects—which ripple around us, as Andreas Gursky has captured so beautifully in photographs like *99 Cent.* The counterparts to these shrines are found in the photos of Edward Burtynsky, who shows us the factories in which these copies are produced, the warehouses in which they are stored before heading off to the malls, and the garbage dumps in which they end up and from which they are sometimes recycled.[1]

This is the other side of copia as abundance—what Taussig, following Horkheimer and Adorno, calls "the organization of mimesis"[2] through the global capitalist economy, the nation-state, and its various appendages and substrates. We might also call it "the modern appropriation of copia," since it is a very particular enframing of copia as universal abundance, as plenitude.

Why are objects stacked en masse in a supermarket or store? As I walk the aisles and look up and down the shelves, what I see are copies, identical copies. When I walk through the supermarket, where bright lights are reflecting off the shiny packaging and the products themselves are hanging in groups, I am sent into a trance. Perhaps, as

Brion Gysin suggests, we all recognize the trace of infinity that is there in the rows upon rows of objects sitting on those shelves. They can look beautiful, as those Gursky photos show. Or rather: sublime, in the sense of something overwhelming that exceeds the senses' ability to take it in, as though the whole of the global marketplace were somehow embodied in those aisles, with their apparently endless waves of consumer products looming above us.

Despite Plato's hostility to mimesis, the world of industrialized mass production of copies appears as a bizarre realization of Platonic philosophy concerning the object. The essence remains an idea which can be implemented not once but an almost infinite number of times, each iteration of the object having the same relation to essence, or lack thereof, as all the others. Processes such as standardization function as a perverse implementation of Platonic idealism, since they encourage the notion that if each industrially produced object is identical to every other, this must be because they are all "perfect copies" and thus stand in undistorted relation to the ideal which they are a manifestation of. Furthermore, if outward appearance is a way in which the idea comes to presence, as Heidegger suggested in his reading of Plato, then identical packaging, seductive presentation, act to confuse us into thinking that we are getting something more real than real itself—not merely the already-distorted copy of the craftsman, but an object that, through the effacement of its own production, can be perceived (falsely, of course) as being somehow closer to the idea itself.

The counterpoint to these supermarkets and mass-produced items can be found in composer/artist Phill Niblock's beautiful movie *The Movement of People Working,* which projects images of people around the world at work simultaneously on multiple screens, to a soundtrack of the vast pulsating drones which Niblock has explored throughout his musical career. This juxtaposition is an enigmatic one, without any obvious explanation. But the droning provides a powerful counterpoint to the repetitive, cyclical, sensuous movements of dockworkers and others, their slow work of transformation of their worlds. The droning constitutes a hub of sameness, but a sameness that, over time, changes in its aspects—as all sustained tones do, and as "people working" do. The discipline, concentration, and mastery involved in sensuous labor are presented sympathetically, but without sentimentality, embedded in the limited economy of a particular marketplace, as well as in the general economy of universal flux.

The Mass-Produced Object

Mass production did not begin with European and American industrialization, or for that matter with Gutenberg's printing press. Any consideration of mass production, even as an object of human consciousness, must begin by recognizing it as a physical and biological phenomenon that is evident everywhere to us: the falling of rain and snowflakes, the growth of leaves on trees, the spreading of trees into a forest, the massing of birds in flight. Reproduction, in the visible world of insects, mammals, and plants, as well as in the invisible-to-the-naked-eye world of microorganisms, occurs mostly through a proliferation of apparently identical organisms, seeds, spores. Even the stars appear as a mass phenomenon, from which humans make differential figurations of various names and forms.

The manufacture of a mass of more or less identical or "standardized" objects and forms by human beings can be traced back as far as the Neanderthal age, in which we know, for example, that humans made such things as beads. But this raises the question: What exactly do we mean by "mass-produced object"? Can a mass of lotus seeds, of the kind used for *malas* in India even today, be viewed as so many "copies"? Are they mass-produced? Copia, as goddess of the harvest, is the goddess of masses, of the massing and gathering of nature, which works through the production of multiples, seeds, leaves, fields of plants, the teeming of tiny fishes in rivers, the endless rolling of waves. But some further act of appropriation or transformation—call it labor, work, production—must be performed on those nature-produced masses before we refer to them as "copies." Thus, Copia is also the goddess of the storehouse, where abundance is measured and held in reserve for the future.

The question of the origin of mass production is an open one, but it is clear that most of the elements have been in use for a long time. Seals and stamps, which are able to mark objects in a repetitive, identical fashion have a long history as markers of property or identity, dating back to at least the fourth millennium B.C. Religious objects, such as the clay figurines known in Tibet as *tsa tsa,* were produced en masse in both Asia and Europe for millennia. These were "copies" of deities brought to life by cultures that believed the deity could find its way into presence an infinite number of times in an infinite number of objects—not because of an idea, but because of rituals of supplication, blessing by religious teachers, infusion with relics, and other tactile mechanisms of mimetic magic. According to Lothar Ledderose, mechanical duplication of bronze vessels in China dates back to the fifth century B.C., and Chinese factories utilized the division and specialization of labor to mass-produce lacquer and bronze objects in the first century A.D. Amphorae were mass-produced in ancient Egypt, Greece, and Rome for transport of wine and other goods. They were often stamped or otherwise marked to identify their place of origin, for otherwise these products would fall back into the sea of undifferentiation.[3]

Looked at from this point of view, the various histories of mass production, in ancient China, Greece, Rome, Mesopotamia; the revival and transfiguration of these methods in the Renaissance, with the printing press, and the modular production of ships and other items in Venice; the Industrial Revolution in England in the eighteenth century; the Fordist assembly line at the beginning of the twentieth century—all can be viewed as chapters in an enormous universal history of the drive to make copies, to mass-produce identical items, and to expand this mass production to every possible sphere by all available means. Above and against all the avowed utilitarian goals that this mass production is said to enable—from the satisfaction of economic wants at a lower cost, to the equitable distribution of needed items—lies a fascination with the magic by which things, including money, can be multiplied, a still-mysterious power. Contagion, named by Taussig (following Frazer) as one of the two components of mimetic magic, is itself a power of multiplication and proliferation; for the unstable act or event by which a mimetic reshaping takes place, whose simplest form is doubling or a single repeat, already implies the possibility of a reoccurring, a repetition ad infinitum.

The fear of being inundated, submerged in a crowd, or trapped in an unending repetition is a basic human fear, although it has taken particular forms in modern societies, where the fragility of bourgeois individualism runs up against the ominous masses who loom, waiting to stake their claim on history. Repetition, as Freud told us, can be the mechanism of repression, but also, simultaneously, of the insistence of the traumatic fact, in the form of symptom.

Mass production reigns supreme today—not just the mass production of consumer items, but the mass production of natural resources secured as a standing reserve to be sold and consumed, the mass production of weapons, the mass production of information in various media. There is also the mass production of markets, both physically in the growth of shopping malls and similar retail spaces around the world, and in the ideology of the capitalist marketplace as the only game in town (aside from the military and the priesthood, both intensely mimetic formations as well) in the age of globalization—a process that is itself a particular implementation of strategies of mimetic installation.

Towns, too, can be mass-produced from a blueprint, as attested by the endless proliferation of mock-Tudor pseudo-villages, houses all built at the same time within a certain margin of variation, in the suburbs of major metropolitan areas around the world. In these towns we find the same stores, the same entertainment centers, as corporations seeking global saturation of markets compulsively spew out copies of their products and business models, with minor variations for whatever remains of local condition—in imitation of the Darwinian model of nature, which is the reigning dogma today. It is an ideology of saturation, of proliferation that continues until an externally imposed limit is met. Even the ideology of individuality and/or uniqueness is mass-produced, through websites like MySpace or businesses such as Dell's build-on-demand computer company, which automate the production of individual identities and products as possible iterations of a predetermined set of options and possibilities that can be mass-produced.

Nations are likewise copies, assembled out of mimetic desire, the desire to be what others around them claim to be, this thing called a "nation"; or conversely, the object of a desire on the part of others which forces a group or space to submit to the installation of the structures of nationhood. Power functions through imitation. In Bertrand Russell's definition, it is the ability to produce intended effects. Power, then, is always mimetic, since the effect is the repetition of the intention. Power works through installing effects and making them endure. Thus the monumental architectures of power, from feudal castles to modern government buildings like the White House in Washington or the Hall of the People in Beijing; but also the everyday rituals of interpellation that constitute modern life—for example, business attire (men in dark-gray suits, sober ties, close-cropped hair, smiling and shaking hands with each other, in imitation of the CEO or national leader), or Louis Vuitton bags as fashion accessories. Countless men and women imitate these images of power.

It is banal but nonetheless true to say that our hunger for copies threatens to consume the world, without our even being aware of what it is we are hungering for. But mass production can also be a progressive force that makes it possible for many more people to have access to things they want or need. We also fear it, because it breaks the taboos on copying that I described in Chapter 3. Mass production reminds us of that teeming biological mass that we come from and live in; and it contains an echo of that greater similarity which we are part of, and the limits to our own separateness and individuality.

Commodity Fetishism

In what, then, does the specificity of capitalist mass production consist? Copia, the goddess of abundance, has been appropriated in the modern, capitalist, industrialized marketplace through what Marx termed "commodity fetishism." The shift from the craftsman's making of a table to the stepping forth of that table as a commodity is a shift in the table's mimetic qualities. Before, the craftsman's table was a reasonably well-behaved Platonic object, the idea, coming to presence in wood, but also maintaining the nature of wood itself. After, the table-as-commodity becomes "something transcendent," "stands on its head, and evolves out of its wooden brain grotesque ideas." It also stands the Platonic hierarchy of idea-and-object on its head, reversing it so that the object now invents ideas.[4] "Fetishism" refers to the practice, supposedly confined to the non-Western world, of attributing powers and agency to inanimate objects. "Commodity fetishism," then, involves discovering a whole set of mimetic magical powers latent within the Platonic object, and setting them free—so long as they conform to the rules of the marketplace. The table has been transformed; it means something different from what it formerly did; it has different powers, "metaphysical subtleties."

What are these powers? First, it has exchange value—in other words, a price—by which it is linked to everything else that has a price. And this exchange value is an abstraction of the labor that went into the production of the object. But as Taussig points out, it is this absorption of the sensuous powers and energy of labor into the commodity that provides the tactile, contagious component of mimetic magic: "The swallowing-up of contact we might say, by its copy, is what ensures the animation of the latter, its power to straddle us." The commodity "conceals in its innermost being not only the mysteries of the socially constructed nature of value and price, but also all its particulate sensuosity—and this subtle interaction of sensuous perceptibility and imperceptibility accounts for the fetish quality, the animism and spiritual glow of commodities."[5]

What, then, accounts for the fetish quality of noncapitalist commodity forms? Arjun Appadurai has argued that the structure of the commodity is not unique to industrial capitalist societies, and has shown that around the world, "things" manifest as commodities according to a variable political organization of exchange and exchange value that includes barter, gift giving, and gift receiving, as well as money-based transactions.[6] We could extend Appadurai's argument with respect to the Marxian fetish too. All forms of fabrication endow their products with fetish powers that contain within them the sensuous work of production, whether this happens through nature, artisanal methods of craftsmanship, factory assembly lines, or the fully automated factory. Speaking more broadly, all actions that go toward shaping particular names, forms, and identities are fetishistic in their transformation and appropriation of the object as object. In other words; whenever something is labeled an "object," the structure of the fetish is already there. Bruno Latour coined the term "factish" in order to draw attention to this process, even in the objective production of scientific knowledge.

The difference between fetishes in traditional societies, where all matter is animated, and fetishes in modern capitalist societies, where all matter has exchange value, at first appears to be one of volume. But as I argued in the chapter on copia, folk cultures always possess the means to "infinitize" their local environments. The abstraction of the products of this local process of infinitization, as they are exchanged in a vast global marketplace, results in a particular mode of fetishization. The folk object or commodity integrates within its outward appearance the marks of environment and history which constitute what Benjamin called its "aura." The capitalist commodity whose outward appearance takes the form of the plastic-wrapped, colorfully packaged object for sale at Walmart is "new"; and by an abstraction of the sensuous labor that went into making it, an evasion of the signs of history that would constitute aura, it is presented as something close to the Platonic idea of the object. As a copy, rather than a thing made imperfect by the wear-and-tear of the world, it appears perversely more close to being ideal. Capitalist commodities present themselves as "perfect copies," meaning the embodiment of the ideal form in an object protected from history and the world.

Money

Money was arguably one of the earliest mass-produced objects. Money appears to have emerged in Mesopotamia and Egypt when agricultural products such as grain and livestock became standard measures of value beyond their actual use value. Among the earliest physical forms of money were objects, such as cowrie shells, that were somewhat rare (cowries were found only in the Maldive Islands), naturally multiple, visually splendorous and thus ornamental and/or prestigious, and that were also invested with exchange value. Coins evolved from the use of particular weights of silver or gold as currency in Mesopotamia, as far back as 3000 B.C. The widespread practice of counterfeiting such weights (by mixing other metals in) led to the Lydians' development of the stamped coin around 600 B.C.; the shape of the coin and the mark stamped onto it guaranteed authenticity and weight.[7] Although the forms of weighed precious-metal coins may have been identical, it is the gesture of stamping each coin that makes coins copies. Such guarantees operate within particular sets of limits, and within a socioeconomic-political structure that assumes a monopoly in the production of coins; if anyone at all could make a copy of a coin, it would cease to have the value that it does. Given that money began as a quantity of precious metals, measured as a weight, and later became coinage whose form symbolically denoted a quantitative value, we might ask: Is there a copying that is purely quantitative and does not involve a symbolic or formal act of imitation? Conversely, can there ever be a purely quantitative entity whose appearance does not depend on an act of mimetic figuration? The importance today of stock indexes, tickertape, and the rest of the iconography of the global financial system is evident—as though even in an age of computerized numerical calculations, some act of figuration is necessary.

Money literally embodies many of the qualities we have discussed regarding the infinite plasticity of mimesis. Money, as the marker of exchange value, the privileged register of economic circulation, is a powerful manifestation of "mimetic energy," with all the supple flexibility of form and value that implies. This flexibility is intimately linked to the possibility of exchange itself, and to the transfer of power and value from object to object, and from person to person. The very qualities that modern money is said to possess—storage of value, abstraction, and convertibility, for example—illustrate the plasticity of mimesis, the incredible ease with which likeness or equivalence can be produced. In this sense, it's no coincidence that "plastic" (as in, "I'll put it on plastic") is vernacular for "money" in its potent but temporary abstraction—the monthly credit card bill being all too real.

A certain tendency of money to multiply is also related to mimesis. Inflation, for example, is often the result of excessive exploitation of the possibility of mass-producing money, i.e., copying banknotes; leverage is a way of mass-producing money through a multiplication of debt. From Girard, we are aware of the tendency of mimetic energy to multiply itself, proliferate; and these dangers are certainly apparent in the history of money and economies, to the point where we might wonder if Bataille's "accursed share," the tendency toward excess that he saw as a universal law, is itself connected to a quality of mimesis.

Branding

Mass production today is not just the mass production of the identical, à la *Modern Times*, although the Walmarts, Home Depots, and Ikeas of this world still exist to service the need for this (along with so-called hackers, in the case of Ikea, who work at transforming the generic items on sale). This is modular mass production in which the mass-produced items are "unique objects," limited editions, customized, personalized, individualized objects, featuring add-ons, deluxe options, and so on. Examples range from the variety of coffee options at Starbucks, to Dell's build-on-demand model of computer manufacture, to Nike's design-your-own sneaker salons.

What holds this mass of options and singularities within a particular set of names and forms is a series of strategies, such as branding, advertising, and marketing, which use mimetic magic in particular ways to transform objects that are essentially generic into highly charged objects of desire. Naomi Klein's *No Logo* provides a good description of the factories where sneakers and other brand-name clothes are made.[8] She shows us how a generic object—let's say, a shoe that fits a human foot and that is made of a rubber sole and a stitched-leather or canvas upper—gets turned into a branded product. The factory workers in Sumatra who make a particular brand of shoe are working with copies of a shoe design to mass-produce an object. But the sportswear company then transforms these copies by adding logos and other design elements, by naming the shoes, by introducing permutable options, and by linking them to a whole advertising and marketing apparatus that transforms the set of associations connected to the shoes.[9]

Branding works because of the same paradox regarding the copy that I set out in Chapter 1. It is the nonexistence of a Platonic essence to the things of this world that paradoxically allows their transformation through mimetic magic into a branded product, but the continuing belief in this essence then serves to help make the products of this transformation appear as natural or truly existing. In his recent book *Lovemarks*, Kevin Roberts, CEO of the advertising firm Saatchi and Saatchi, talks about the enormous emotional investment and charge which is the true currency of successful branding. Indeed, the possibility of transfer of ungrounded patterns of energy, which have the ability to attain being in temporary, multiple ways through techniques of mimetic magic, is the basis of branding and advertising. This transfer is usually accomplished through an intense feeling such as desire or fear, which, as we all know, are emotions that can be easily manipulated.

Although we are continuously immersed in the rhetoric of individuality, free choice, and uniqueness, which branding exploits in a variety of ways, there would be no possibility of creating a product line, let alone a brand, without the serial chains of similarity that allow for identities to emerge out of a mass of copies. In the words of Jean Baudrillard: "The serial nature of the most mundane of everyday objects, as of the most transcendent of rarities, is what nourishes the relationship of ownership and the possibility of passionate play: without seriality no such play would be conceivable, hence no possession—and hence, too, properly speaking, no object. A truly unique, absolute object, an object such that it has no antecedents and is in no way dispersed in some series or other—such an object is unthinkable."[10]

Without this seriality, there is no such thing as a brand. Shell Oil could decide to diversify into breakfast cereal, and Merck pharmaceuticals could start its own record label, but it would be difficult to do this without some historical, productive trace that facilitated this. The attraction of Louis Vuitton bags is that the original objects themselves look like exquisite copies. The very idea of branding is contingent on a company's ability to separate the existential objects that it produces from the idea or image of them. Branding is always a revaluation and an appropriation, and the very thing that it insists on, the uniqueness of the brand, is necessarily impossible, or, to use Baudrillard's word, "unthinkable." Of course, there is no reason this insight should not also be appropriated. On the Web, one can purchase bags emblazoned with the legend: "This is a fake LV bag." But surely it is only a matter of time until Vuitton markets its own fakes, designed by Damien Hirst or Jeff Koons.

Compression and Amplification

The seriality of objects around us today is obvious to all, but branding is only one of the ways in which it is produced. Another way is through scale—for example, Starbucks' Tall and Grande coffees, a crafty appropriation of branding into the language of scale, increasing the identificatory power and prestige of Starbucks' product. Many brands today produce items whose value is tied to the scale on which they are produced: limited-edition runs of items manufactured in a different color or slightly different style, such as Adidas' Missy Elliott Bass Line shoe; or mass-produced versions of *haute couture* items, such as Armani's AX line. The meaning of the product is intimately linked to knowledge of the number of copies that exist.

Manipulation of scale is one of the basic mimetic strategies. Canadian multimedia artist David Rokeby uses surveillance and digital-scanning technologies to explore the power of compression and amplification.[11] The simplest example of compression would be the movement from hearing a concert in a concert hall, to listening to a CD recording of the concert, to listening to an MP3 of the same concert. Each step of the process entails considerable compression. In the first case, the full complexity of the concert performance (mediated by amplification) is translated and compressed into digital data, which is considerably less rich than the original sound. This data is then compressed tenfold further, into an MP3 file. A similar process happens visually when one takes a photograph of a landscape, prints from the negative, digitally scans the print into a TIFF file, and then converts the TIFF into a JPEG. Rokeby points out that at every stage in the process, technicians make decisions as to what parts of the original they need to preserve in order to maintain the similarity of the compressed file to the original, and which parts are unnecessary and can be discarded. When something is compressed, the compression is a creative act of transformation of the original, and the consequences of this shift of scale are considerable. My iPod, about the size of a deck of cards, contains the equivalent of several rooms full of vinyl LPs or a sizable wall of CDs. It compresses a mass down to a tiny size.

Amplification is also a powerful tool for transformation of an object. Not only can one stretch the object by making it larger, so that the bits are distributed over a larger area of time and space, but one can also build larger, high-resolution versions of the object, whether it be the vast statue of Maitreya that is currently being built in Bodhgaya, full of sacred relics, or Jeff Koons's gigantic poodle at the Guggenheim in Bilbao. An elegant digital example would be Leif Inge's *Nine Beet Stretch,* which time-stretches a digital recording of Beethoven's Ninth Symphony so that it lasts twenty-four hours, adjusting the pitches so that they match those of the original, creating dense, glacially shifting walls of sound that still contain the melodic and harmonic qualities of Beethoven's original, in nearly unrecognizable form.

All decisions as to scale are creative ones. Such decisions are a basic form of copia, and the production of difference within the same. To quote Morton Feldman, one of the first composers to devote himself to the exploration of repetition:

> Like that small Turkish "tile" rug, it is Rothko's scale that removes any argument over the proportions of one area to another, or over its degree of symmetry or asymmetry. The sum of the parts does not equal the whole; rather scale is discovered and contained as an image. It is not form that floats the painting, but Rothko's finding that particular scale which suspends all proportions in equilibrium. . . . For me, stasis, scale and pattern have put the whole question of symmetry and asymmetry in abeyance.[12]

In Feldman's work, "scale" means the number of repetitions of a melodic shape in a particular piece, and thus also the duration of the piece. Feldman's *Second String Quartet* runs for close to six hours of variations on a chord-like cluster of notes.

In a variety of Buddhist devotional practices, the use of symmetrical and other scalable elements makes possible the creation of forms which manifest a specific and recognizable "likeness" or pattern that expands or contracts according to the situation. One of the foundational texts of Mahayana Buddhism, the *Prajñāpāramitā Sutra,* exists as the *Prajñāpāramitā Sutra in One Syllable* and also as the *Prajñāpāramitā Sutra in 100,000 Lines*—and these texts are not "different." A mandala or "mind palace," providing a blueprint for the visualization of the universe as an arrangement of interdependent enlightened forces and elements, can exist as a narrative text, a 2D painting of almost any size (e.g., on canvas, in sand), a 3D sculpture, a mental image, or a variety of other things. The symmetrical quality of mandalas suggests the way that pattern can emerge from an essenceless, mirrorlike, groundless repetition in which one side repeats the other. Although such symmetry is something we take for granted—in the shape of our own bodies, in nature, in our various productions—it underlies much of what we find both disturbing and fascinating about copying.

Many Buddhist practices involve repetitions and visualizations of phenomena on a vast scale. The Ngöndro or preliminary practices of Tibetan Buddhism, for example, involve one hundred thousand prostrations, the same number of recitals of the hundred-syllable Vajrasattva mantra, and so on. Other Buddhist and non-Buddhist schools have similar quantitative practices, the goal of which is to saturate the individual and the universe simultaneously with a particular relative, cognitive structure, to point to and produce through practice and repetition a recognition of our always already-existing absorption in the sphere of nonduality. The mass production of brands, commodities, saturation advertising, propaganda attempt a similar level of saturation, but with no other aim than the monopolization of consciousness for purposes of control.

Fountain(s)

> Art has been confounded with the art object—the stone, the canvas, the paint—and has been valued because, like the mystic experience, it was supposed to be unique. Marcel Duchamp was, no doubt, the first to recognize an element of the infinite in the Ready Made—our industrial objects manufactured in "infinite" series.
>
> —Brion Gysin, "Dream Machine," in *Back in No time: The Brion Gysin Reader*, ed. Jason Weiss (2001)

In 1917, one R. Mutt of Philadelphia (a.k.a. Marcel Duchamp) submitted a found object, a "readymade," for the first exhibition of the American Society of Independent Artists. The object, entitled *Foun-*

tain and signed on one edge, was a lavatory urinal, turned on its side and placed on a pedestal. The exhibition jury rejected the object, claiming: "It is, by no definition, a work of art."[13] The object sat behind a partition for the duration of the exhibition, after which it was sold and then lost. Today, we know *Fountain* primarily through a photograph of it taken by Alfred Stieglitz. But during his life, Duchamp also made several copies of the piece, consisting either of found objects like the original, or of replicas cast in full size or in miniature.

The urinal used in *Fountain* was an industrially produced object, one of a series manufactured from a mold, en masse, by J. L. Mott Ironworks of Fifth Avenue, New York. Placed on a pedestal in an art gallery, the object became singular—a copy presented as an original, although its creator was nevertheless accused of "plagiarism" and lack of originality.

Duchamp theorized *Fountain* and his other readymades through the concept of the infrathin: "The difference / (dimensional) between / 2 mass produced objects / [from the same mold] / is an infra thin / when the maximum(?) / precision is / . . . obtained."[14] The infrathin establishes a minimal unit of difference which, conversely, establishes the absolute singularity of all objects—including apparently identical mass-produced artifacts, such as urinals produced from a single mold. Thus, Duchamp observes: "The possible is / an infra-thin— / The possibility of several / tubes of color / becoming a Seurat is / the concrete "explanation" / of the possible as infra / thin."[15] Even the most "perfect" copies are different, because their spatial situation and thus their relationship to their environment must be different—they cannot be identical. Also, they cannot be composed of exactly the same physical matter—the molecules of which they are made are not the same. Furthermore, the time and place of their production must be slightly different, even if both objects were created in a mass-production facility using the same mold—again, because a machine cannot produce two objects in the same place at the same time; it produces them either sequentially in time in the same location, or simultaneously but adjacently. Finally, all objects, mass-produced or not, have their own unique histories—as did *Fountain,* made from a urinal that Duchamp had bought used—and it is these unique histories which produce what Walter Benjamin called the "auras" of objects.[16] While Benjamin argued that mechanical reproduction destroyed the unique aura of objects, and while few people are interested in the extremely subtle and delicate set of distinctions that would mark off one particular mass-produced object from another, there is a great deal of importance to these distinctions. Any object, whether naturally occurring, handcrafted, or factory made, is unique in that it is composed of unique physical matter, occupies a unique point in the space-time continuum, and has a unique passage through that continuum, meaning that it has a unique history. In this sense, all objects can be said to possess an aura; and phenomenologically and otherwise, this is what it means to say that "this object exists." And here we are not even thinking about another issue that interested Duchamp: whether an object is the same object it was one second before (from the point of view of physics, it is not).[17] In other words, whether the being-in-time of all entities and objects has as its correlative a singularity or uniqueness that manifests at every moment within nonduality.

The gesture of drawing attention to an infrathin is commonplace in contemporary art, where everything from a drugstore (Damien Hirst) to classic paintings and photographs (Cindy Sherman, Sherrie Levine), to pop-cultural imagery and text (Andy Warhol etc.), to a pile of bricks has been re-presented in a gallery or museum context. But the infrathin is also operative in folk cultures—in the repetition of generic motifs and devices such as particular songs, rhythms, patterns, and practices, in situations where the singularity that is evoked is not merely a singularity wrested from the illusory appearance of the identical, but a significant, contingent, affectively potent singularity.

Duchamp himself was rather cautious about readymades, and in his 1961 essay on them, he notes: "I realized very soon the danger of repeating indiscriminately this form of expression and decided to limit the production of 'readymades' to a small number yearly. I was aware at this time that for the spectator, even more than for the artist, art is a habit-forming drug, and I wanted to protect my 'readymades' against such contamination."[18] Again, we come up against the danger of a contagious proliferation of objects that begins as soon as a mimetic process is initiated. The comparison of these dangers to a drug recalls Plato's mimetic pharmakon. Repetition is "habit forming"—it can lead to Hegel's bad quantitative infinity, where "one more" is added an infinite number of times.

The Information Object

Mass production today increasingly means the mass production of digital objects. The imagined proximity of the digital copy to Plato's ideal form is radically changing our relation to objects, so that the actual object is undergoing a major devaluation in favor of virtual objects, which form an increasingly "loud cloud" around us. My iPod has 12,990 songs on it, including duplicates. Like Benjamin with his book collection, I have not listened to most of them, and I know they exist only because they appear within the grid of a database on iTunes in a way that locates them.

One of the basic uses of computers is to give names to objects in the form of particular strings of ones and zeros. In his groundbreaking 1979 story "True Names," computer scientist Vernor Vinge spelled out many of the possibilities for creating multiple identities, for surveillance, and for tracking that have become basic facts of twenty-first-century life. "True names" are the names by which you can be tracked, the digital code that underlies the Dungeons and Dragons mythical names that the protagonist and his proto-chatroom denizens use, as well as the everyday names which hide the hacker's true identity and agency. In his book *Shaping Things*, science-fiction writer Bruce Sterling describes what he calls the "Internet of things." The history of object tracking in business goes back a long way, and is most familiar to us in the form of the barcode—the set of black and white stripes, attached to most industrially produced objects, that a scanner can read and translate into information which can then be connected to a computer database. Sterling describes the recent development of arphids (a.k.a. RFIDs)—digital tags, with a small radio attached, that can broadcast the identity of the object to systems capable of tracking the object in time and space. These arphids become the names of objects, since "naming enables the generation of pattern. Naming enables measurement."[19] They are not generic, as a barcode is, but unique to each object.

What Sterling means by the "Internet of things" is the possibility of being able to search for and track down any object in the world using its arphid identity. According to Sterling, the result "is that I no longer inventory my possessions inside my own head. They're inventoried through an automagical inventory voodoo, work done far beneath my notice by a host of machines. I no longer bother to remember where I put things" (93). Furthermore, because every object is in some sense at hand—can be called forth on the Internet of things (Heidegger would say that *where* it is no longer matters, since all distances are similar, but *what* it is is no longer known, since what is closest to us remains a mystery)—the actual material presence of the object is unimportant. Indeed, "at many other times, many crucial times of serious decision, I'm much better served with a representation of that object" (95). The object itself is now "merely hard copy" (96) which can be made or found when required, while a "weightless, conceptual interactive model that I can rotate inside a screen"

(95)—in other words, the digital file—holds the true identity of the object, including the history of its real-life uses.

This is the object as pattern, as information, as accumulation of ones and zeros. It is then concretized by a machine called a fabricator that spews out copies of the information object known as "fabjects." From Sterling's point of view, this system will make possible a true ecology of matter and material objects, which will be designed and tracked from idea to store to consumer to wastebasket to recycling heap. Thus, what appears to be a joke at Plato's expense, with "ideal form" replaced by "information object," actually ends up reflecting Plato's caution with and suspicion of material objects—i.e., fabjects as mere false copies of the real ideal forms.

In the light of what I have said about copia, there are reasons to doubt this ecologically correct view of copying. If industrialization came about because a few wealthy people were able to get their hands on technologies that allowed the infinite replication of material objects, a capability they then proceeded to indulge to extreme excess, the democratization of material industrial production through the universal availability of "fabricators" is likely to cause an exponential increase in material production—as the ubiquity of personal computers and printers has resulted in a rapid growth in paper consumption and printing.[20] The production of copies, as I hope I have demonstrated already, is a matter of *passion,* of almost bottomless fascination, and the idea that the workers of this world will enjoy it any less than their bourgeois predecessors is questionable.

While the mass production of objects has been shifted to non-Western countries, Western economies concern themselves with the creation of services, lifestyles, information. The folk-cultural practice of copia provides a set of models, styles, and blueprints that can be appropriated from the community that developed them and mass-produced somewhere on the planet, then marketed and sold globally.

But one of the consequences of the personal computer is the possibility that, whereas formerly mass production was limited to a small elite class of bourgeois entrepreneurs, today many more people can mass-produce copies of things. And this is the major crisis concerning copying today. Terrorism is spoken of as "asymmetrical" warfare because the size of the threat is much larger than the quantity of participants, wealth, or conventionally defined power involved. Computer-virus attacks, denial-of-service attacks, spam, the distribution of MP3s through peer-to-peer networks, news blogs, and the like all make possible an asymmetrical production and distribution of copies with considerable consequences. Different phenomena have different levels of resistance to being copied and reproduced in this way: music, itself so close to nonduality, is eminently copiable and lends itself well to mass distribution, whether by phonograph, cassette, CD, or MP3. Matter, including human beings, is much more difficult to digitize, although back in the 1950s Norbert Wiener argued that teleporting humans was by no means impossible, just technically very difficult for now.[21]

Digital Perfection?

If there is no such thing as a perfect copy in the material world, can we say the same thing about a digital copy? Mathematically, it seems possible that two copies of a computer file could well be composed of completely identical data—i.e., the same stream of ones and zeros. At first glance, these two copies of a file would be identical in a way that two of Duchamp's urinals could never be. They might genuinely be identical. But there are objections to this at a number of levels.

First, from a physical point of view, it is as impossible for two copies of a digital file to be stored in the same place as it is for two copies of an object to occupy the same point in space-time. The place where digitally copied things are held, waiting to be called up, replicated, recombined, is a database or filing system, which is where a digital copy is physically located, in the form of a series of ones and zeros, symbolically registered as bits of high and low voltages stored in a series of compartments. Thus, just as two urinals cannot occupy the same space, the same is true for two digital files, which would have to be stored either on different computers or in different locations on the same hard drive. Even if they were on identical hard drives, the way in which they were stored would be different, since files are not stored as a linear sequence. The two copies that were generated from another copy would also have a certain error rate, which, although quite low, means that files of any significant size would not be exactly the same. And again, when the code was executed, each file would be executed either on a different computer in a different digital environment from the other, or on the same computer but not simultaneously, and so the execution of the code would be different, both in the time elapsed and in the outcome. The difference might be small, but the files would not be identical.

New-media theorist Julian Dibbell has pointed out that the question of how a one and a zero are constituted in different computing systems is also indeterminate—that ones and zeros are not absolute quantities, but simply temporary differences in voltage. In a recent conversation, he observed:

> We tend to think of bits as these sort of atomic, on-or-off monads, but they are usually represented as two different voltage levels—1 being thus-and-such a voltage, 0 being another. And since there's usually a gap between the levels, and large numbers of electrons involved in determining a given voltage, there's lots of room for physical difference at the electron level between two digitally equivalent bits. Digital information really is nothing more or less than a form of writing. Just about any question you ask about bits can be illuminated by asking it about script, I find. So: "How different can two electronically coded 1's be?" is sort of like

asking how different two 1's written on a page can be. The answer to the latter is: very different indeed, as the disciplines of typography, calligraphy, and handwriting analysis attest. The "invisibility" of electronic code makes it sort of opaque to these disciplines. But is it impossible to imagine that there might one day be a sort of calligraphy of the bit?[22]

Thus, Duchamp's infrathin, the smallest possible difference between similar things, asserts its full power in the computer: the difference between a one and a zero is an infrathin, or is becoming an infrathin as technologies seek the smallest, the briefest, the most subtle measurable difference between states of energy or matter—the infrathin as the minimum, minimal unit of information. Stretched out in ever-longer chains of ones and zeros, the power of the infrathin is maximized. Going back to Walter Benjamin's dictum, explored in Chapter 1, concerning the interplay of sensuous (i.e., semiotic) and nonsensuous similarity in mimesis, we can say that ones and zeros are semiotically different, but that this difference is reduced to the minimum needed to be technically perceptible. The digital realm relies on the fact that energy itself is discontinuous. The mimetic power of the computer—the flexibility with which it establishes new kinds of mimetic relationships between entities—may be related to the proximity of sensuous and nonsensuous similarities. In other words, digital ones and zeros represented as tiny differentials of energy are as close to being "the same" as it is possible to be, while maintaining a quantifiable difference. And they point to, or figure, that nonconceptual mimetic stuff/nonstuff through which we fabricate the relative world of name and form which we inhabit.

The Politics of the Infrathin and the Infinite

We vacillate continuously. We are all the same; we are not the same. We are copies produced by the shuffling of genetic code; we are all unique individuals with our own specificity and contingency. As Henry Flynt points out, many of the claims that are made to the effect that we are all identical machines are mere posturing, and the all-too-human actions and affects of those who make claims of being robots or genetic clones or networked computers reveal themselves as such at every moment.[23] At the same time, the creative appropriation of images and discourses of uniformity is a key part of contemporary culture, from drag queens and kings to the Crips-and-Bloods gang aesthetic in hip-hop, to the various mutations of skinhead. All speak of a knowing embrace of copying as a strategy that produces a rule, a game, a challenge out of the tension between same and different.

But is the difference between man and man not also an infrathin? Freud coined the term "narcissism of small differences" to describe the exaggeration of this infrathin as a way of trying to establish individual separateness and self. But the acceptable margin of our differentness is also very small; corporeal and mental deviations from the norm remain disturbing to us, and, in the past, excision from the community could be absolute. Conversely, excessive similarity is also considered disturbing. As control over the most delicate and obscure areas of consciousness and embodiment becomes a concern of the various operative forms of government, and as zones of normality and pathology are defined in ever-finer detail, we are increasingly called upon to present ourselves as copies, as repetitions of a certain model, a certain framework. And this is no longer enforced through a crude behaviorism, as classic modern dystopian texts like *A Clockwork Orange* or *1984* imagined, but through drugs like Prozac, the manipulations of genetic engineering, or the requirements of the job market. Conversely, mass production today involves the demand that we present ourselves through the production of differences, through acts of mimetic transformation that can be framed within the order of the marketplace. The same / not the same.

It would be a mistake to conflate the general phenomenon of

mass production, particularly in its folk forms, with industrial capitalism. Indeed, the gap between the two has been usefully deployed in key moments of struggle against imperialism—for example, during Gandhi's campaign for *swaraj* (self-rule) in India, which included the boycotting of industrially produced cloth and clothing from Britain. Gandhi, controversially, went further in an attempted refusal of even domestic industrial production. He proposed that Satyagraha Ashram members, and (later) Congress members, and (later still) all Indians make their own *khadi* cloth, as a move toward autonomy and self-determination *(swadeshi).* Early attempts to produce cloth at the Satyagraha Ashram floundered, due to lack of knowledge and technology, until Gandhi met Gangaben Majmundar, who offered to locate the traditional *charka* spinning wheel and teach him how to use it. After Gandhi learned the skill, the ashram began a program of spinning practice and production which all ashram members were required to participate in. It also promoted the distribution of copies of the *charka,* education projects in *khadi* production, and outreach programs to village communities.[24]

To put it within the framework that I have elaborated over the past few chapters: Gandhi mobilized a folk technology for the mass production of cloth, in opposition to an industrial mass production that imitated it. The technology itself could be duplicated and disseminated throughout Indian villages and cities, in the form of an object (the *charka,* or spinning wheel) and a set of practices. More important, the practice no doubt already existed in many places. But Gandhi arguably made some mistakes in attempting to disseminate a particular mode of making cloth among diverse communities, whose members already used a variety of local and traditional cloth-making practices—methods that, Gandhi argued, they should give up. This was the modern error of attempting to universalize, or rather nationalize, a particular folk modality as symbol of the people, rather than supporting the development and flourishing of a variety of folk forms, including those that appropriate industrial technologies.

We are left with the problem set out in Phill Niblock's movie—the one with which we began this chapter: What is the universal that articulates the desire and demand of the diversity of moving bodies, without turning them into the mechanized zombies of *Modern Times?* Perhaps it could only be a sound, unstruck or otherwise, like Niblock's droning, or like the glossolalic song that Chaplin sings toward the end of his otherwise silent movie. While music remains framed as entertainment within our existing societies, or the kitsch universality of "We Are the World" and other globalizing abominations, it has also emerged, through hip-hop and other forms of electronic dance music, as one of the principal vectors of an actually existing *autre-mondialisation,* the revealing of a "planet of drums (or rather drum machines)," in Steve Goodman's recent formulation, which transforms Mike Davis' "planet of slums."[25] Profoundly at home in the uncanniness of repetition, and the particular powers of digital sound manipulation, popular Afrofuturist dance musics have proliferated in the first decade of the twenty-first century, from kuduro, to dubstep, to coupé décalé, reggaeton, and cumbia—all diasporic but developed in specific locations. To twist Bob Dylan's definition of folk music, these are unconstitutional reruns of mass production. Neither modern nor not-modern, they work with industrial capitalist commodity forms, but are committed to other forms of mass production. Goodman refers to such subcultures as mobilizing a "bass materialism" which, he says, "is enacted as the microrhythmic production and occupation of space-times by collectively engineered vibration" (172). As Goodman recognizes, the politics (or "subpolitics") of such collectivities remains an open question, yet it offers a powerful example of a liberatory mass production that is happening today.

7/Copying as Appropriation

Property relations in Mickey Mouse cartoons: here we see for the first time that it is possible to have one's own arm, even one's own body, stolen.

The route taken by Mickey Mouse is more like that of a file in an office than it is like that of a marathon runner.

—Walter Benjamin, "Mickey Mouse" (1931), trans. Rodney Livingstone, in Benjamin, *Selected Writings*, vol. 2 (1999)

What if Appropriation [*Ereignis*]—no one knows when or how—were to become an insight whose illuminating lightning flash enters into what is and what is taken to be? What if Appropriation, by its entry, were to remove everything that is in present being from its subjection to a commandeering order and bring it back into its own?

—Martin Heidegger, "The Way to Language" (1959), trans. Peter D. Hertz, in Heidegger, *On the Way to Language* (1971)

Everything Is Appropriated

I remember the first time I taught my class at York University on copying, the week we came to discuss appropriation, plagiarism, and the like. I gave students my definition of "appropriation"—the act of claiming the right to use, make, or own something that someone else claims in the same way. Thinking about appropriation enables us to ask: Who has the right to make a copy? Which people have the right to prohibit someone else from copying them or that which they believe belongs to them? A student raised her hand and said that if this was the definition, then the slave trade had to be considered a vast act of appropriation. There was a lovely, sad silence in the room; and after a second, I responded that most of what we call history is arguably the history of appropriation, and the history of one group stealing from another group and claiming those people's bodies, minds, properties, lands, or cultures as their own. This history continues today unabated, and it brings up the philosophically complex problem of belonging. While we must acknowledge the importance of the juridico-political discourse whose role it is to decide questions of belonging, and the trauma that accompanies what is called "theft," if we want to understand what is at stake in speaking of "copyright" and the controversies that accompany it, we must ask: What can we truly say belongs to us? To what degree have we genuinely given consent to the structures and situations in which we find ourselves, including those that establish what "belongs," and to whom? And if we look broadly at human history: What is there that has not been appropriated by others in the name of some idea or entity or structure? And finally: In what sense do identity and essence *ever* really belong?

There is a long history of appropriation in the arts. To take a few lines of an author's composition, to copy an image or a melody and use it in your own work: such acts of citation or outright theft formed the basis of art before Romanticism—Shakespeare's extensive use of other playwrights' plots and texts, for example. The valorization of the expressive power of the individual artist emerged around the same time as copyright laws, during the Romantic period. But the integration of the original artist into the marketplace was also accompanied by the rise of an avant-garde whose work has constantly been built around a critique of notions of originality, identity, and property. Such avant-garde work includes collage and montage by Picasso and the Dadaists; direct acts of appropriation such as Duchamp's *LHOOQ,* a retitled and retouched print of the *Mona Lisa;* Warhol's soup cans and silk screens. More recently, there was the movement known as "appropriation art," which launched the careers of artists such as Cindy Sherman, Sherrie Levine, Jeff Koons, and Richard Prince, not to mention writers such as Kathy Acker in New York in the late 1970s. As Nicolas Bourriaud notes, in today's "postproduction" art world, appropriation as the recycling of circulating images and forms is a basic strategy—Damien Hirst, for example, has moved a whole pharmacy from store to gallery. The impasse of appropriation in art was described accurately by Benjamin Buchloh twenty-five years ago: every calculated act of transgressive appropriation made by experimental artists today speculatively assumes a future recuperative appropriation into art history and the culture of the museum, making transgression a shrewd investment.[1] This impasse also describes a broader crisis of the copy today, which includes question marks appended to "work," "identity," "ownership," and "community."

Appropriation is about a lot more than a particular artistic strategy, and in our attempts to reveal the broader context of copying, appropriation has always been our theme. Counterfeit Louis Vuitton bags are appropriated in the sense that the idea of the LV bag, its outward appearance, and its signs are produced inappropriately by those who lack the legal right to do so. But an act of appropriation was likewise carried out when the family-run company called Louis Vuitton, *malletier,* was absorbed into the conglomerate LVMH in a more or less hostile takeover. Plato's description of mimesis in the *Republic* is in fact a description of appropriation, since the imitating artist takes the *eidos,* or outward appearance, from one place and produces it in another, appropriating the form or materials. From the emergence of Copia out of the goddess Ops to the display of plastinated dead bodies in Gunther von Hagens' *Body Worlds* exhibitions, everything that we have discussed has involved an act or event of appropriation.[2] And it is in this broader history of appropriation that I seek to find a way out of the current impasse.

Theft as a Universal Principle

> Mutual stealing among the three powers makes everything appropriate for its time.
>
> —Zhang Boduan, "Essay on Achieving Perfection," quoted in *Daoism and Ecology*, ed. N. J. Girardot et al. (2001)

What is "appropriation"? The word and its cognates have at least two contradictory but related meanings. First is the sense in which the noun "appropriation" is used above: the act of taking something and making or claiming it as one's own, or using it as if it were one's own. Second, the adjective "appropriate" denotes that which is proper to a situation or a person, that which is "appropriate." Appropriation often involves taking something which arguably belongs to someone else. There is the sense of seizing, of making a claim on something that is claimed by someone else, of stealing. The adjective "appropriate" refers to that which one has a right to claim as one's own, that which is "properly" one's own. The term is thus intimately related to the concepts of property, ownership, and rights.

Appropriation, in the sense that is familiar to us, occurs when there is a dispute as to ownership of an object, or the image or reproduction of that object. Slavery is an extreme example: the act of appropriating a human being, turning that human being into an object, into property, renaming that being, forcing him or her to do certain things, to speak the language of the master, claiming the right to define and use that being over and against that being's own self-determination. A slave, in a very real sense, is a copy of a human being, with all the lack of respect traditionally accorded to "copies of things," rather than "things themselves." Even in the absence of slavery *qua* slavery, the history of human societies is a history in which certain individuals claim power over others, and the form this claim takes is that of a series of appropriations. History, on this view, is the history of appropriations, of events of appropriation, of the endless chain of pacts, exchanges, territorial claims which fill conventional history books.

Lest the notion of a universal history of slavery functioning through appropriation seem an extraordinarily gloomy idea, I would like to look at a Taoist view on these questions, one that points to the kind of ecology of copying that I have discussed in previous chapters. In his "Essay on Achieving Perfection," Taoist master Zhang Boduan (who lived around 1000 A.D., during the Sung Dynasty) argues that theft is actually a universal principle, since all beings exist and survive by stealing from each other.[3] One could counter this argument by speaking of sharing, but the history of the world is not solely a history of sharing. The kind of theft that Zhang Boduan describes is universal, for nobody gets out of here alive, and nobody gets to take anything when leaving. At a basic level, the molecules of the universe are named and claimed by different groups of beings or forces for various periods of time; but everything from an amoeba to a president to a star eventually dies, losing name and form and the right to associate the two (which is the principle of identity). Biologically speaking, we appropriate oxygen and nutrients from the time we begin to develop as separate entities during embryonic development. Our bodies are constantly being appropriated, too (10 percent of our dry weight consists of bacteria, according to biologists).[4] The language we use to describe our many acts of creative copying—words such as "influence" (inflowing), "inspiration" (breathing in), "absorption" and "digesting" (eating), and "incorporation"—are all metaphors for constitutive processes of autopoietic (i.e., homeostatic self-regulating) appropriation. Their corporeal nature suggests their importance, as well as our need to situate this importance in one of those gray zones between nature and culture, where physical necessity meets mental habit.

Appropriation is also an important practice in folk cultures, where it is related to the use of guile and cunning as a strategy of the powerless. Stealing is integral to many myths of the origin of mankind—for example, the myth of Prometheus (telling of his theft of fire from the gods) and its many variants in shamanic and trickster stories. It is also integral to many stories concerning the origin of particular folk practices. For example, the Chinese internal martial art known as Yang-style Tai Chi Chuan is said to have begun when Yang Lu-Shann traveled to Henan province to study with the great master Chen Chang-Shen. Because he was not part of the family, he was treated badly and kept at a distance. One night, however, he was awakened by the sound of someone practicing in a nearby building. He got up and saw his master practicing the secret techniques, and night after night he returned to watch and secretly learn this practice, which he studied with great devotion. His mastery of the technique made him invincible during practice, and, seeing this, Chen Chang-Shen finally gave in and taught him everything.[5] Such stories help us to resolve one of the paradoxes of traditional folk cultures: How, with their concern for family and tribal secrecy, do "secrets" that form the basis of generic regional folk practices become disseminated? Such cultures maintain a kind of open secrecy in which what we might call "intellectual property" is guarded and controlled on the one hand, but acquired through revelation or trickery on the other.

Property and the Open Secret of Universal Appropriation

The universality of appropriation is an open secret—known to everyone but almost impossible to speak of. We should remember this when we are talking about "cultures of copying" or, for that matter, about copyright law disputes concerning digital reproduction of music. Ultimately, there is only emptiness, the nameless, wild, chaotic nonduality that lies beyond all concepts and labels, where "matter" is in a continuous state of self-devouring flux and where forms live, mutate, and die from second to second. In the Tibetan tantric tradition, this is known as *durtrö,* literally "charnel ground" or "cremation ground," but metaphorically the space of impermanence and transformation.[6] The human world, where I speak of myself, am given a name, and enter a specific society, is a particular, highly contingent iteration of this universe, and we go to great lengths to sustain our image of permanence as it constantly runs up against the facts of impermanence, dependent origination, and emptiness. We do so through acts of appropriation that provisionally and temporarily "give" us name and form. In Buddhist philosophy, the Sanskrit word *upādāna* can be translated as "appropriation," and the word is given an almost exclusively negative meaning, as one of the twelve links of dependent origination which make up samsaric or worldly existence.

Any discussion of property, intellectual or otherwise, must begin with this basic recognition: in any society, what I consider to be mine can be taken away—because ultimately nothing is mine, nothing belongs to me, and finally there is no me. But it does not inherently belong to others, either. Law as an institution exists as a way of resolving issues around appropriation—it attempts to set out the rights of individuals and institutions in naming, making use of, and owning things. But the law itself appropriates in order to do this, because the power to enforce the law must be taken from someplace, fabricated and consolidated in a forceful but nevertheless impermanent way.

The most important and celebrated Western philosophical examinations of property acknowledge the fundamental nature of appropriation.[7] In John Locke's theory of property, which provided the philosophical basis for the first formulations of copyright law in Britain in the eighteenth century, property is appropriated from nature through labor. Ownership, says Locke, begins with our bodies, and their capacity for labor and work. Through the sensuousness of labor, man establishes ownership of the commons of nature and God: "His labour hath taken it out of the hands of nature, where it was common, and belonged equally to all her children, and hath thereby appropriated it to himself."[8] Although Locke does not say so, ownership is established mimetically: the contagiousness of the conceptual "me" and "mine" passes through "my" work on the world around me, allowing me to appropriate elements of that world.

There is a rich body of nineteenth-century writing on appropriation and property—in particular, French anarchist philosopher Pierre-Joseph Proudhon's *What Is Property?* (1840), which contains the famous statement "Property is theft!" and which led to Proudhon's correspondence with Marx. That the 1800s should have produced such works is hardly surprising, since at that moment the European powers were engaged in appropriation on an unprecedented, global scale, turning the whole earth into a web of private property. The young Marx, no doubt influenced by Hegel, pushes the concept of appropriation considerably further than Locke.

> Man appropriates his all-sided essence in an all-sided way, as a total man. Every one of his human relations to the world, seeing hearing, smelling, tasting, feeling, observing, sensing, willing, acting, loving, in short, all the organs of his individuality . . . are, in their objective relation *(Verhalten),* or in their relation to the object, the appropriation of it. The appropriation of human actuality, its relation to the object, is the exercise of human actuality, human activity and passivity or suffering.[9]

The term "appropriation" occurs repeatedly in Marx's work, most famously in the *Communist Manifesto,* which is rich in relevant discussion: the appropriations of the feudal system and the bourgeoisie are to be ended in favor of an appropriation of private property to the proletariat through a worldwide communist revolution. But the problem or question of appropriation itself is barely addressed by Marx and Engels. They simply assume that appropriation as such is necessary and justified, and that it is merely the direction of the appropriation that needs to change. In the twentieth century, we saw the results of this refusal to address appropriation as a problem. "Real existing communism" degenerated into the appropriation of individual and collective wealth by the state, by dictators or other groups claiming to represent the proletariat. This is the impasse of appropriation as it exists today: we have no convincing or believable model of appropriation beyond the Lockean one.

Expanded concepts of "the commons," "the public domain," and "fair use," which inform progressive opinion on legal reform of IP law, all rely on the affirmation of a space in which mutual appropriation is sanctioned, valued, and encouraged. But such initiatives, however admirable, are always belated, for contemporary cultures of copying are, as I have shown, summoning before our eyes a different kind of object and commons that, though avowedly novel and unfamiliar, we have known about for a very long time. Knowledge of copia in this sense was suppressed during the industrial age through the ability of certain groups, in both communist and capitalist societies, to monopolize the production of copies, and to enforce copyright and intellectual-property laws that disenfranchised individuals and collectivities. Subaltern, subcultural, and other marginal folk groups are scapegoated by global capitalism for criminal acts of reappropriation, but global capital is itself nothing other than institutionalized and legitimated appropriation on a vast scale. As a Somali pirate pointed out in a recent interview with a journalist, "We don't consider ourselves sea bandits. We consider sea bandits those who illegally fish in our seas and dump waste in our seas."[10]

The irony is that Locke's theory of property, which of course is open to question in a variety of ways, can support a variety of social structures, since the tactile mimesis of sensuous labor describes everything from the movements of hunter-gatherers to industrial communes, and, by the logic of that ownership-conferring tactile mimesis, the alienation of labor under capitalism would be either impossible or illegal. Bernard Edelman explains the ideological maneuver which ensures why this is not the case. He tracks the debates in early history of photography and cinema, where it was argued that the photographer or cinematographers should not be called creators, and could not legally own the images that they made, because it was the camera rather than they themselves that was doing the work. Similarly, it could be argued that the worker in the factory is not entitled to appropriate the fruits of his or her labor, since it is in fact the industrial machine *qua* capital which is doing the work, and that machine belongs to the owner.[11] Of course, as photography and cinema became a part of industrial society, the photographer and filmmaker were retroactively accorded status as artist and creator, along with the rights of appropriation, "since the relations of production will demand it" (49). Edelman goes further, claiming that "all production is the production of a subject, meaning by subject the category in which labour designates all man's production as production of private property" (52).

Comparative study of property regimes reveals enormous variety in the structures by which humans develop notions of property. But clearly, this is not just a human question, for, as the poet Gary Snyder reveals so beautifully in his essay collection *The Practice of the Wild*, the natural world is full of structures and practices that necessarily negotiate the problem of appropriation and appropriateness. Appropriation, therefore, is a matter of ecology, specifically an "ecology without nature," as Timothy Morton recently put it.[12]

The Politics of Appropriation

The question of who gets to appropriate is a fundamental one. Spike Lee's movie *Bamboozled* offers us an intense look at the appropriation of blackness, and the paradoxes of identities that are simultaneously enforced with the crudest violence, appropriated at will for fun and profit, and evaporating in a postmodern haze. The movie starts with a song that situates the events of the movie after "1492" and the Middle Passage, two colossal acts of appropriation. The film's protagonist, Pierre Delacroix, is a TV-show writer. He's a parody of a "buppie," a black urban professional with a slightly pedantic Harvard accent that marks him within the movie as one of V. S. Naipaul's "mimic men," imitating whiteness. In response to the threats and taunts of his white boss, Dunwitty (who thinks the fact that he's married to a black woman gives him the right to use the word "nigger" and hip-hop slang), Delacroix proposes that the network do a minstrel show, based on the activities of two young black men who tapdance for spare change on the street outside the TV company's offices. To Delacroix's surprise, the network is wildly enthusiastic about the show, and the movie shows the consequences of their decision to broadcast *ManTan: The New Millennium Minstrel Show.* Messy and confused, but with wisdom and cinematic insight, *Bamboozled* explores a cast of characters who all appropriate blackness in different ways; and as the movie unfolds, the politics of this appropriation become increasingly unclear. The strident Dead Prez–style politics of the "blak" hip-hop crew, the Mau Maus, are treated as caustically as the politics of Delacroix's astute but alcoholic comedian father or his striving but sexually compromised assistant, Lamb.

Although Lee's overt message is that misrepresentations of blackness such as blackface, minstrelsy, and the posturings of gangsta and political hip-hop are morally wrong, Lee and his camera are clearly fascinated by the mimetic power of blackface, both in archival clips of old Hollywood films and in Delacroix's young protagonists on the new minstrel show. And everyone laughs at the minstrel show—humor crackles contagiously through the whole film. Although the black characters are clearly motivated to adopt certain kinds of appropriation in order to survive, the motivation of the white characters such as Dunwitty, or the white members of the TV audience who look nervously to the blacks in the audience before responding to the minstrel show, is less obvious. When push comes to shove, they can produce their whiteness like a credit card. And when they try to repudiate it, this whiteness is appropriated to them by force, as when the only white member of the Mau Maus is carried away by the security forces who have gunned down his nonwhite partners, despite his pleading to be allowed to be executed along with them.

As we have seen, some people are allowed to transform and some are not. In his essay on imitation in colonial situations, Homi Bhabha argues that imitation is forced on colonial subjects so that they may appear as "proper" and functional servants of empire, but it is also limited or bounded so that the colonial subject is never allowed the colonialist's full status of being. The "proper" political subject undergoes processes of mimetic transformation that endow him or her with the status of citizenship and legal-political identity. The mimesis of the colonial subject, however, always "fails"; it is demanded but at the same time repudiated, ensuring that those who are governed but who lack rights are thrown back into the inauthenticity of the mere copy, empty of essence. Thus, the distinctions and power that hold together structures of inequality are maintained.[13] The problem is not that mimesis is bad or wrong, but that the freedoms of open, unobstructed mimetic transformation which are enjoyed by those who have power are denied to the governed. The latter are forced to obey a highly ideological framing of mimesis, whose political uses were already set out by Plato in the *Republic.*

The most surprising moment in *Bamboozled* occurs late in the proceedings. The show has proven to be a wild but controversial success and Delacroix is sitting in his office, which he has filled with kitsch—racist knick-knacks such as a Nigger Piggy Bank statuette featuring an effigy of a black butler with big lips and bulging eyes.

As Delacroix ruminates on his success and guilt, the statuette rolls its eyes and grimaces. The other statues, "fetish" objects in various senses of the word, come alive and seem to state their disapproval of what Delacroix has wrought. It is as though these tawdry, kitschy, distorted copies have suddenly become possessed by angry ancestor spirits and come back to life.

Many of the issues in the film resonate with broader issues concerning cultural appropriation that have repeatedly been raised in recent years with regard to indigenous and subaltern groups.[14] Coco Fusco's essay "Who's Doin' the Twist? Notes towards a Politics of Appropriation," for example, points out the way that the celebration of an aesthetics of appropriation, avant-garde or otherwise, tends to overlook underlying issues of power and privilege which determine who is able to do what.[15] Although Fusco tries to separate "cultural exchange" from "appropriation," she offers no criteria for doing so beyond the predictable demand for an analysis of particular situations. I am unpersuaded by her suggestion—or Lee's—that marginalized peoples appropriate out of necessity, whereas privileged peoples do so as an exercise of their power.[16] The act of hiding, disguising, or naturalizing appropriation serves to support a particular power structure (based on domination), while the exposure of the appropriation can potentially produce more flexible but no less potent power structures also built around appropriation. When Dunwitty says, "If the truth be told, I probably know 'niggers' better than you, Monsieur Delacroix. Please don't get offended by my use of the quote-unquote N-word. I got a black wife and three biracial children, so I feel I have a right to use that word,"[17] his claim is a complex blend of self-exposure and defensive hyperbolic appropriation. Finally, as the name of the character suggests, it's the crass stupidity of Dunwitty's appropriations, including his crude deployment of his family situation for the sake of his career, that is the problem.

But is it impossible to imagine someone in Dunwitty's situation who makes transethnic or transracial identifications in a way that is proper, progressive, or even radical? Isn't there a conservatism in Lee's portrayal of mimesis that ends up scapegoating all mimetic activity in the name of an authenticity whose existence can never be proven, but only summoned up as a set of fragments of the past in the form of copies? Lee himself was accused of similar acts of unethical appropriation and inauthenticity by Amiri Baraka in connection with Lee's film about Malcolm X. But again, the struggle there is not between a true, rightful, natural owner of the legacy of Malcolm X and an Other who is delegitimated, but between two acts, events, or claims of appropriation that are of undeniable significance but that come with "no guarantees," as Gayatri Spivak says.[18]

While traditional societies maintained a variety of complex property regimes and relations, the translation of these rights into private property that can be defended within the structures of existing intellectual-property law is merely a stopgap measure—at least if the goal of this translation is the protection of the vitality of a traditional culture, rather than the enrichment of certain members of the group who are able to appropriate the common wealth as their own. Appropriation should be affirmed not only as something done to such cultures, but as a vital and dynamic part of their own self-constitution. Perhaps more important: insofar as such cultures functioned through practices that are critical of appropriation and implicitly resistant to the capitalist structure of private property, their practices need to be celebrated and developed as components of an invigorated and prospective commons.[19]

Ereignis

One of the epigraphs to this chapter is a passage by Heidegger, from a lecture given in January 1959. In it the word "appropriation" translates the German word *Ereignis.* The latter is conventionally translated as "event"; but in consultation with Heidegger, Joan Stambaugh used "appropriation," even though this word is more commonly rendered as *Aneignung*. Others have translated *Ereignis* as "event of appropriation," "being on the way," or "enowning." Still others say it is untranslatable. As early as 1919, Heidegger used *Ereignis* to talk of the way we receive the gift of Being, how it is appropriated in order that the world we live in appear. In his second, esoteric book, *Beiträge zur Philosophie* (Contributions to Philosophy), written in 1936–1938 and subtitled *Vom Ereignis* (Of Appropriation), the term took on a more specific meaning: an event in which a nation or people finds its destiny. After World War II, Heidegger modified this claim in the light of the National Socialist ideology that it appears to support; but throughout the latter part of his career, he continued to emphasize the significance of appropriation, going so far as to claim, in "The Way to Language," that it was "richer than any conceivable definition of Being."[20]

Heidegger uses the concept of *Ereignis* in working through his critique of the Platonic doctrine of identity-as-essence. Appropriation, in the sense of seizing something that belongs to others and making it one's own, belongs to the tradition of metaphysics, since it posits that things have essences that belong to them, and at the same time that these essences can be stolen. The paradox is that if these essences can be stolen, they can't really be essences—a transferable essence being a contradiction in terms. For Heidegger, the process by which things come to appear to have essences relies on an appropriation; in other words, the essences which appear to belong to them are appropriated to them. Thus, it is appropriation, rather than essence, that is determinative of these things.

But there is an ambiguity in the epigraph I've chosen from "The Way to Language," since it is unclear whether, when things are removed from their subjection to the commandeering order (of metaphysics, science, or industry), they will be "restored" to their own order or to the order of appropriation itself. What could "own" imply, if not an essence? Where does "properness" come from, if not essence? In his late works, Heidegger spoke of "dwelling" as relating to a home that was properly one's own but nonetheless constructed, and of nearness or proximity, rather than essence, as a measure of Being. Nevertheless, the question of how any sense of belonging can be constituted, phenomenologically or otherwise, remains unresolved in his work—and has remained an important topic in contemporary philosophy all the way through to Alain Badiou.

The major objection to the notion of essencelessness that I have put forth as a way of understanding copying is that if nothing has an essence, then surely anything can be anything, anything could belong to anyone, and nothing could belong to anyone ever—in which case, why or how could anyone ever make any claim to legitimate ownership of anything, even his or her own name or body? One could say that what we call "property" is merely the consequence of thuggish enforcement strategies propagated by a gang of robber barons who have the power to enforce their claims and, through brute force, establish them as law even when they are illegitimate or false; but although this notion has some appeal, it also has major weaknesses. Our common experience of the world is not complete and arbitrary chaos, and when we look at a tree, although we can recognize the impermanence of the tree—the fact that it grows, changes with the seasons, eventually dies—we nevertheless feel that a tree is not a bird or an ocean, nor could we call an ocean a copy of a tree (or could we?). The tree is established by an act of labeling that is an appropriation, yet we also recognize a certain properness in calling a tree a tree.

Heidegger was interested in these two simultaneous qualities of appropriation, qualities that seem to contradict each other. If the word "tree" appropriates the object labeled as such, where would any notion of "properness" in the designation of "tree" come from? Heidegger's most persuasive response to this problem was the concept of gathering and the "four-fold": the mutual appropriations of earth, gods, sky, and mortals which establish entities in the world.[21]

The properness that relates the word to the object, which establishes properness and even property, is a quality of "being-with," of dwelling, of closeness and constellation—which is to say, a relationship that is sustained over a period of time.[22] It is not an essence that establishes properness, since it is clear that the doctrine of essences, as the long history of post-Heideggerian critique tells us, is ideological and serves to legitimate particular appropriations, while presenting them as natural facts. The doctrine of "being-with" is no less political, but it is a politics in which the construction of claims about identity, about what "is," can be evaluated clearly and openly, even if the clash of claims may itself be very complex, weaving historical and genealogical and scientific and religious threads together. This dense weave, however, *is* what we call "subjectivity" and "objectivity," and we pass over it at our peril.

No doubt copies, especially punk-rock MP3s and appropriated surveillance-camera footage, are among the many things, modern or otherwise, which Heidegger would have excluded from his definition of "Being." But the theories of the four-fold, of dwelling, of intimate proximity are all mimetic theories, and they apply not only to activities of copying and objects labeled "copies," but also to originals and to the path-wandering philosophers who find them. Modern technology, which Heidegger describes in explicitly mimetic terms, was something to be repudiated; yet when he traced its "essence," he also considered it part of the "saving power." Doesn't all appropriation—indeed, all copying—have a technical structure, and act potentially as a revealer of Being? While remaining fully aware of the reifications of "the folk" made by the National Socialists and others in the nineteenth and twentieth centuries, couldn't we say that Heidegger is actually articulating a politics of folk cultures in the broader way I have described them? "Folk cultures" defined as those who maintain that practices of copying have the power to reveal Being and to fabricate community?

Appropriation and/or Depropriation

"To appropriate," in everyday parlance, means to take something so that it belongs to us, whether it be the likeness of something or the object itself. But "to appropriate" can also mean to put something in the place where it belongs, or—from the other side of such an event—to submit to or allow that act by which we are placed in the place where we belong ("to be appropriated"). All copying, seen from this point of view, is an appropriation because it takes the outward appearance of one thing and brings it forth in another.

Let's suppose that the "someday" when "appropriation" would bring everything back to itself is now upon us. It is unclear that Heidegger would recognize that this was so. Nevertheless, much of the most profound thought concerning the earth and humans since World War II, whether coming from ecologists or performance artists, computer software developers or poets, philosophers or political activists, has struggled to respond to the challenge of appropriation, and to find new ways of understanding belonging that go beyond the ideas of work, mimesis obtained through likeness, contagious tactility, right, and so on.

Is appropriation avoidable? Both Nietzsche and Marx thought not. In a remarkable passage in the *Grundrisse,* Marx argued:

> All production is appropriation of nature on the part of an individual within and through a specific form of society. In this sense it is a tautology to say that property (appropriation) is a precondition of production. But it is altogether ridiculous to leap from that to a specific form of property, e.g. private property. (Which further and equally presupposes an antithetical form: *non-property.*) History shows rather common property (e.g. in India, among the Slavs, the early Celts, etc.) to be the more original form . . . But that there can be no production and hence no soci-

> ety where some form of property does not exist is a tautology. An appropriation which does not make something into property is a *contradictio in subjecto.*[23]

There is a certain pathos in the final sentence of the quote, as though the thought that there might indeed be "an appropriation which does not make anything into property" had occurred to Marx, regardless of whether it made any sense. The failure of existing communist societies in the twentieth century might come down to an inability or refusal to think *Aneignung* as *Ereignis*—to go beyond merely reversing and repeating the appropriation of the commons from the bourgeoisie to the ruling party.

Nietzsche argued that "life itself is essentially appropriation, injury, overpowering of the strange and weaker, suppression, severity, imposition of one's own forms, incorporation and, at the least and mildest, exploitation. . . . It is a consequence of the intrinsic will to power which is precisely the will of life."[24] A certain history of the avant-gardes, of the failings of the countercultures as hierarchical, male-dominated groups of the most traditional kind, can also be located here in the impasse of appropriation. It is quite possible to read Nietzsche's final mental collapse in Turin while watching a horse being flogged in the marketplace as the manifestation of a crisis in his faith in the will to power as the fundamental right of appropriation.

In a text addressing his former comrades in the Popular Front, Bataille opposed appropriation to excretion—excretion as the inevitable loss of that which has been accumulated, as that which must be disposed of.[25] He argued that any concept of revolution that consisted solely of a transfer of rights of appropriation was inadequate, and that the real revolutionary project was that of organizing extravagant projects of waste disposal that go beyond the traditional methods of scapegoating, war, etc. This remains our problem today, and it suggests profound reasons why there are limits to any legal resolution of problems of copying. Bataille's solution, it should be noted, like Heidegger's, was also a reinvigoration of traditional and folk practices of ritual within the modern world—a solution that was treated with blank incomprehension while he was alive. But the "problem" of what to do with an inappropriable excess also informs our most intimate activities. Even the apparently self-evident notion that our bodies are our property falls apart when one examines the complex appropriations and counter-appropriations of lovemaking for example. Making love means letting go—it could not happen without the relinquishing of ownership, done passionately, elegantly, or otherwise.

Again: Is appropriation avoidable? The question is an important one in thinking about copying, because the unappropriated or unappropriable object or event would be a peculiar kind of thing / no-thing which would suggest different ways of responding to and inhabiting the world, and a different relation to likeness, to appearance, and to play.

Heidegger's use of the word *Ereignis* rather than the more conventional *Aneignung* indicates his own caution concerning the appropriateness of the appropriate, and the direction of appropriation. English and French theorists addressing this area have written of "propriation," "reappropriation," "misappropriation," "ex-appropriation," and "depropriation," as well as the "unappropriable." The thought of appropriation is intimately linked to that of the gift, and of the "donation of Being" as the ultimate horizon of the intelligible and sensible worlds. When the phenomenal world is given to us, should we take it and make it ours? Can we refuse it? If it is not given, should we steal it? This is a mimetic problem, since to say "mine" or "ours" is already to enframe, to take, to copy.

What would it mean to speak of a "depropriated" subject, object, or copy? Consider the following lines from Hélène Cixous' "Laugh of the Medusa": "If there is a 'propriety of woman,' it is paradoxically her capacity to depropriate unselfishly: body without end, without appendage, without principal 'parts.'"[26] "Depropriation" here relates to the maternal-feminine that I have discussed and the ability to allow things to happen—feelings, process-events such as childbirth, relationships of various kinds—without imposing a set of heavy-handed conditions governing self, other, essence, identity as preconditions or control mechanisms organizing, regulating, or otherwise governing the movement or flux of being. In the realm of the maternal-feminine, there is no phallus, no thing, that marks or attempts to center identity. "She doesn't lord it over her body or her desire," Cixous continues. The sculptures of Louise Bourgeois or Shary Boyle embody this flux of limbs and forms, simultaneous additions and subtractions that dissolve normalized, appropriate, gendered conceptions of what a body is, while maintaining a keenly material or corporeal orientation that is mimetic without any deference to Platonic ideas of what a body should be.

"Depropriation" here means indifference to possession. It indicates a willingness to relate to the world without imposing conditions of ownership in doing so, an ethics of care that does not require ownership, that requires an ethos other than that of ownership in order for there to be caring. It means allowing to circulate according to context, and therefore to remove from the logic of appropriation, and from enslavement to a particular context that is naturalized as "what must be." Depropriation is a form of "renunciation," in the Mahayana Buddhist sense of the word, where what is renounced is not the object but attachment to and fear of the object, and the acts of labeling that these relations to the object involve. In this sense, "depropriation" could mean liberation; and although my interpretation here is not an orthodox Buddhist one, depropriation could be understood as the reversal of *upādāna,* and of appropriation as one of the twelve steps of dependent origination.

Buddhist literature is full of stories that involve depropriation. For example, the famous story of the Indian Mahayana master Asanga, who prays for a vision of Maitreya, the future Buddha, for many years, without success, until one day he finds an old dog lying by the side of the road, covered with wounds that are infested with maggots. Overcome with compassion for the dog, he starts to remove the maggots, but then realizes that he will likely harm the maggots by doing so. He therefore places his tongue in the wounds and invites the maggots to climb onto him. Whereupon the dog transforms into Maitreya.

Depropriation sounds scary, and it is easy to distort or vulgarize the concept in order to discredit it. The word has been used to describe certain kinds of violence—for example, the committing of murders and disappearances in places like Colombia, where not only is the victim taken from his or her village at night and killed, but the victim's face and other markers of identity are destroyed. Certain torture strategies also involve the erosion of the victim's identity and the assumed ownership of his or her body and mind. But they also involve very aggressive strategies of reappropriation: the goal of torture is to make the mind of the victim the possession of the torturer, to make it a thing. The goal of the disappearance/murder is both to erase an enemy and to send a warning to others, in order to dominate a group of people or an area. French mystic, philosopher, and political activist Simone Weil placed her practice of "decreation," one which bears remarkable similarities to depropriation, in opposition to "destruction," and argued that the ethics of decreation consisted in uprooting oneself, one's ego, voluntarily and with discipline, rather than uprooting others.[27]

Can depropriation and what we call "copying" coexist? Copying is a form of appropriation because making a copy involves positing a relationship between two objects, the name of one being given to another, the form of one being produced or recognized in the other. "To appropriate" means to make a claim of identification and property, in the sense that the claimed object has a name or form that belongs to it. "To depropriate" would mean to let go of such claims, and in this sense it would be to abandon the notion of copying and the control over name and form that the notion involves. But this brings us back to "copia" as examined in Chapter 2, where myriad versions of something, a folk song for example, are passed around a community. Sometimes this object, and the lineage through which it passed to the present user, are acknowledged. At other times, the song is disseminated anonymously and without identification. We recognize the former as copying. But is the latter case, which is surely more prevalent, any less mimetic?

We are continually depropriating and depropriated. One way of understanding nonduality is as depropriated similarity or appearance: free, open, and unobstructed. The depropriated copy would be something like those MP3 files floating around iTunes that have no name or label. Technically, they are copies, yet they have become invisible, silent, without label but nonetheless present. Probably most of the copies in the universe are like this: depropriated, unclaimed, unrecognized, or misidentified. Why bother to call such a thing a "copy"? Because one cannot understand copying without recognizing that the difference between original and copy is merely one of designation, and that both original and copy are ultimately nondual. So, then, perhaps an ocean can be a copy of a tree, or, in the words of Zen master Dōgen, rivers can be mountains.

Improvisation and Used Objects

> That's the great thing about improvisation. Or *playing*—"improvisation" has got that heavy sound to it. Playing is really subversive of virtually everything. So you clamp it down, like the industry's clamped down on it. I mean, they don't want improvisation, naturally. You can't make money out of this shit where you don't know what's going to happen from one minute to another. So the process has been, of course, to nail it all down. But then the subversiveness gets into the technology, so

> even a guy doing a mix—you can't nail him down. There are guys improvising remixing a record. And that's where the life is in music. It always seems like it's the vein, the conduit for life in the music. That appetite seems to me to be always to do with changing things, which is often to do with fucking things up.
>
> —Derek Bailey, quoted in Christoph Cox and Daniel Warner, eds., *Audio Culture: Readings in Modern Music* (2004)

> Type ex.—worn trousers and very creased. / (giving a sculptural expression of the individual who wore them) / the act of wearing the trousers, the trouser / wearing is comparable to the hand / making of an original sculpture.
>
> —Marcel Duchamp, *Notes* (1983), note 44

Strategies of depropriation surround us and constitute many of our most significant pleasures and practices. In the arts, the long history of the avant-garde comprises a series of attempts to depropriate the commodity form and its attendant social structures: from Duchamp's *Fountain,* to the Surrealist practice of automatic writing, to Situationist notions of *détournement* (more or less literally, "de-appropriation"), to Cage's and Fluxus' "dematerialization of the art object" (in Lucy Lippard's phrase), to the evolution of happenings and performance art, all of which involved varying degrees of anonymous or collaborative production.[28] For the most part, these depropriations succeeded temporarily, as "events of depropriation" or "temporary autonomous zones" (in Hakim Bey's phrase),[29] but were then appropriated into the marketplace, or into art history. Could it be otherwise? Possibly. But in order to recognize this, one would have to look beyond fine art, even at its most "subversive," toward popular practices of depropriation that are not always appropriated back into the stable categories of the art world.

The practice of improvisation as developed by jazz musicians, and later by Derek Bailey and others, aims at producing an event that can never become a copy, because it is a singular event. The implications of such an improvisation are too numerous to treat adequately here, but they include: the erosion of the line that separates performer and audience, accompanied by a destabilizing of notions of profession and expertise that produces a new (but hardly unknown) type of collective; and the challenge of a "being-with" based on a dynamic, immanent sense of relationship to what's going on. But the problem of the copy is never far away. As Bailey notes, faced with a field of total, open possibility, many improvisers repeat a certain set of gestures that are "free" but as predictable as the idiomatic forms they seek to move away from. In other words, they copy themselves, or they copy a way of relating to other musicians. This is not necessarily bad, since it can result in new idioms, protocols, forms of beauty and pleasure (what Simon Reynolds identifies as the pleasure of "cheesiness").[30] Or not: there are no guarantees.

The dynamic of appropriation and depropriation can also be seen in the open-source software movement, which is concerned with determining a set of rights of appropriation that serve the interests of various communities, as opposed to the prevailing model of copyright and private property that companies like Microsoft embody. Although open-source software has expanded the rights of users to access computer code and change it—and although, relative to prevailing copyright models, it is therefore depropriated—almost all such software does come with a license that defines and structures the possible appropriations of the code. In other words, a radical depropriation of code is quite rare, and there is very little interest in it. The reasons for this are various, but principal among them is the need for some stability of the coding platform to allow the community of programmers (and users, too) to work together. A "copy," in this case a repeatable structure of code, functions as an event in which a community comes into being, provisionally, temporarily. The crude capitalist version of this insight results in the endless platform wars between corporations over DVD formats and the like—attempts to appropriate and control a structure in which "choice" will be offered, subject to various IP restrictions.

But the Internet is perhaps unavoidably a space of appropriation, defined by and as code. There are other, more everyday forms of depropriation. For example, several of my students, when I asked them to write about branded objects, said that they do all their shopping at the Canadian thrift store Value Village, and sang the praises of used clothing, which they argued evaded the logic of ownership. I was surprised at their response, but when I think of Derek Bailey's argument regarding improvisation, I can sec the parallel. Business wants fixed objects—objects that can be mass-produced, which is to say copied over and over again, and then sold many times. Improvisation doesn't produce this stable object. But the power of those who appropriate, those who buy used clothing and so on, comes from the fact that this stable object never really existed in the first place. The "new" object is the object fabricated and appropriated by a business and put on the market for the first time in its current, temporary form. This is a "use" of other, preexisting objects ("raw materials," "natural resources," "chemicals," and other passport-less entities), and all the packaging around "new" objects exists to obscure the fact that the object has already been "used" in the process of becoming a commodity.

It's the changing object—not the objectified one—that's the real object. And "copying" relates not to an unchanging original from which copies can be made, but to the ways in which one relates to an object and subject that are always changing. Shoppers at Value Village may enjoy the marks of an object's history, the ways in which it has been used, the signs that it has been possessed by other people. When they take possession of such objects, they know that this possession is temporary, and that they will hand the object on to others. They live in a phantasmagoria of marks and hauntings which may be more important to them than the "use" to which the object is put. They are improvising with objects, and they are relating to the changing qualities that every object has, and by which it is constituted. They are also relating to their own changing nature, enjoying the temporariness of their own relations with the object.

The other side of used objects is eBay and the like, which Fredric Jameson has wryly called "our collective unconscious."[31] We find, on eBay, a remarkable archive or museum of objects, memories, all for sale, all waiting to have exchange value bestowed upon them, to be reappropriated; but in the meantime, they float in cyberspace like Bruce Sterling's information objects. Museums have traditionally served as markers of a shift in the status of objects in society, and of their appropriation to some new group or system of values. The first major public museum in the world, the Louvre in Paris, opened in 1793 during the French Revolution, making the royal collections of art and objects open and accessible to the general public. And museums continue to be the repositories of colonial plunder, full of ritual or sacred objects torn from their contexts and owners and presented for the curiosity of the public, in the name of an individual, a group, or a state.

Nowadays, whenever anyone makes a powerful observation regarding the nature of objects and the complex powers that come with naming, property, and ownership, it will inevitably be appropriated into the capitalist marketplace to appear in commodity form. Thus the artfully faded new-but-aged jeans sold by companies like Diesel—garments that strive to imitate the *wabi-sabi* of a thrift-store pair of used jeans while being new and unused. But as McKenzie Wark has argued, that marketplace is likewise continually being appropriated in various ways by hackers and others.[32]

Depropriation Means Learning to Relax

> Our poetry now / is the reali-zation that we possess nothing / anything therefore is a delight / (since we do not pos-sess it) and thus need not fear its loss / We need not

> destroy the past; it is gone / at any moment, it might reappear and seem to be and be the present / Would it be a repetition? Only if we thought we / owned it, but since we don't, it is free and so are we.
>
> —John Cage, "Lecture on Nothing," in Cage, *Silence* (1973)

What is most difficult to uphold is the appropriation and/or depropriation that is not respectful, that takes something without asking and uses or changes it. A conventional ethical view would argue that such appropriation is wrong and that it shows a lack of respect for propriety and for a history which would establish the rights of certain people to exclusive or privileged use of something. But it is precisely this kind of appropriation that is prevalent today, and it is also a kind of appropriation that on occasion has the most fortuitous results. Such an appropriation does not obey laws of "cultural exchange," and it is usually asymmetrical. But this doesn't mean it's used solely by the privileged or powerful on the marginalized and powerless, since it's also employed by the marginalized and powerless. In the formulation of Michael Hardt and Antonio Negri, global capital, in its contemporary form of empire, appropriates the wealth of the commons through legal protocols such as patent and copyright registry, by which it establishes ownership. The goal of the multitude—those who are poor because they are denied access to material wealth and the realm of the immaterial that includes ideas, identities—is to reappropriate that wealth.

Folk cultures, which collectively are equivalent to Hardt and Negri's "multitude," are always feeling their way toward situations, events, times (Hakim Bey calls them "temporary autonomous zones") where such reappropriations are possible, and where their full potential, as copia, infinite abundance, can be manifested. What is significant there is not the transgression, since the taboo is imposed by particular sociopolitical regimes. What is significant is the goal—which is the enjoyment of copia, not only as an infinity of objects but also as the free multiplicity of the subject. The inappropriate appropriation or depropriation is linked to the problem of the gift, as Derrida has formulated it. Indeed, insofar as Being itself is an appropriated gift, isn't it always somewhat inappropriate? From Eve's tasting of the apple (the mimetic figuration of knowledge) to Prometheus' theft of fire (the mimetic contagion of power), the role of stealing in myths of origin suggests that this is the case. But if man must steal the property of the gods in order to initiate his Being, where is the gift?

Can a copy be a gift? This question has come up a number of times in this book, from the discussion of the relative merits of mixtapes and CDs as gifts, to the various folk strategies for manifesting copia, to the fabled "gift economy" of the Internet and the fruit seller's courgette for the baby. For Heidegger, appropriation's importance lay in the way that the gift or donation of Being was received: only through the gift's appropriation was there Being. Politics, as Heidegger knew, begins with the possibility of an "original appropriation." But as soon as appropriation occurs, we are already in the territory of mimesis, the supplement, and the copy. Depropriation offers the possibility of a refusal of this logic, without a concomitant refusal to engage with the gift of that which becomes Being. This is not a refusal of mimesis—just a refusal to engage in an appropriative mimesis. Therefore, this gift cannot be entirely separated from mimesis *qua* mimesis.

Nevertheless, the notion of copy-as-gift should strike us as odd. If, as Derrida tells us, the gift transcends all possible economizations, exchanges, and reciprocities, copies are, for most people, viewed as an entirely economic phenomenon.[33] The copy, according to conventional wisdom, has no value of its own; it has value only because of what it imitates. At the same time, it exists solely because the thing it imitates has value and is therefore *worth* copying. But a copy is often thought of as an unsuitable gift because it is cheap or free, and a “worn copy” (to use the title of a recent Ariel Pink record), is just about worthless. But Ariel Pink’s copies, built up from loops of old cassette tapes, are extremely charming precisely because they are worn, just as the mixtape becomes a powerful gift because the signs and traces of difference appear all the more powerfully against the sameness of copying technology.

Copying has the potential to evade the logic of economization and equivalence for the opposite reason that the gift does: if the gift has a value in excess of all conventional valuations, the copy seems to be deficient in value, too trivial to value. It’s “just a copy,” whether a photocopy, an item of used clothing, or yet another free MP3. The copy is never allowed the myth of essence that is accorded to other things, and that is used to establish their value. If the copy has a value, it is established only through deception and dissimulation, through being substituted for that which it imitates, that which it is mistaken for. But it is not the same as that which it imitates, and thus it has no apparently autonomous value. Except that since nothing in the world has an essence, and the establishing of ownership and property through reference to essence is therefore an illusion, the copy itself is potentially closer to the mystery of *śūnyatā*, and to a radical reformulation of what is meant by “value.” Copies are free—free of value, free of identity—and they bring with them the news that everything is virtually free, and could perhaps *actually* be free, too.

You may say that such a vision of a universe of free copies is impossible. A certain impossibility of the copy also links it to the gift. As we have established, there are ultimately no copies, no absolutely identical things; yet we persist in believing that there are, in order that we might *get rid of them*. Similarly, we imagine gifts that are beyond all laws of exchange and economy, in the face of strong evidence that there are no such gifts, believing this in the hope that we might *get to own them*. Such belief helps to account for the extraordinary persistence of the Platonic notion of imitation of essence or ideal form, despite all attempts to deconstruct it—a notion that appears in some form whenever a philosophical system of some kinds takes on the problem of representation and the relative world. Essence is a potent myth, one of the phantasmatic structures that allows mimetic figuration to take place. But perhaps the positing of essence is merely an excuse, an alibi, a rule, as Baudrillard characterized it, that allows a certain type of play to occur, in a way that is compelling in spite of, or in collusion with, the rule of law.

We speak of "abundance," but it would be more accurate to speak of an "essencelessness," whose figurations include an infinite, abundant multiplicity. Just as the gift points toward that emptiness which is the true nature of all phenomena (and which is ultimately neither gift nor not-gift), the abundance of copia points to it too (one translation of *śūnyatā* is "fullness"). Copia may be continually broken up, appropriated, quantified, and labeled as "copies" functioning within a limited economy. No doubt that is why Copia is hard for us to recognize as it is. But with every upload of a music file, every generic pair of shoes that appears in a marketplace frequented by the poor, every martial-arts move that is repeated, every fuck and every prayer, folk cultures do recognize the groundless infinite abundance of Copia, and that is why they persist.

In his essay on mixtapes, Thurston Moore insists that laws will never prevent people from sharing music, because "trying to control sharing through music is like trying to control an affair of the heart—nothing will stop it."[34] Thus, copying, as we have already seen, is connected to love. To reiterate a comment made at the beginning of this book, what I have written here is an affirmation rather than an ethics. Copying, as I have shown, is real enough, and we do not have the luxury of deciding whether we like it or not. The question—in the words of Buddhist poet John Giorno—is how we handle it.[35] Some degree of honesty regarding our participation in and reliance on copying takes us a lot further than a zealous concern for justice and rights. To put it bluntly: in order for there to be something we call a world, containing an "us" and a "them" and an "it," there has to be copying, too. It is in this sense that I think traditional and folk cultures are more realistic than contemporary capitalist society, with its insistence on copyright. Things pass easily from person to person in folk cultures, even though they also have a keen sense of property and propriety. But the trickster god is always nearby, to remind us that life and all the forms it takes are impermanent, constantly changing.

I'm not a libertarian, and I recognize the need for checks and balances that prevent those in power—those who have already appropriated so much—from using their power to appropriate everything else. Respect, care for the particularities of transmission and dissemination, is important; and in this sense, some version of copyright law is "appropriate," as are current attempts to defend and expand the public domain, the commons, and fair use, and to promote "dynamic fair dealing." Particular communities have rights to the production and use of copies associated with them. But in the age of globalization, "copies"—for example, digital media files, pharmaceuticals, political tracts, fashion designs—move fast, and the communities that form around them dissolve even faster. These copies that we share and struggle over make a demand on us that goes much further than previous arguments concerning the rights of subcultures and subaltern cultures to appropriate aspects of dominant cultures and repurpose them, or struggles to imagine an equitable global consumer culture. I believe this demand is in the direction of the depropriated copy—the copy in whose assembly we are able to recognize the interdependence of everything, a recognition that would manifest as the Net of Indra, as the depropriation of the limited subject and object back into the dynamic flux of nonduality, also known as *śūnyatā,* or emptiness.

An understanding of depropriation is integral to most folk cultures. It forms the basis of a commonality, and a set of diverse but mutually supporting practices that negotiate the paradoxes of mimetic transformation in meaningful ways. I have made the argument about folk cultures and depropriation using vocabularies taken from modern philosophy and cultural theory and from Mahayana Buddhist traditions. A Mahayana Buddhist understanding of *śūnyatā* has enabled us to correct weaknesses in the post-Nietzschean theorization of the relation of the different to the same, and to think depropriation in terms of practices that can make possible the recognition of a pervasive nonduality. The various theoretical traditions—Marxist, Freudian, and Nietzschean—also allow us to correct certain weaknesses in Buddhist traditions of thought concerning social formations, power structures, and the politics of mimesis.

The idea of depropriation presents enormous challenges to all of us—whether subaltern or the most privileged citizens of the most economically powerful regimes. What is more threatening than the thought of giving up a possession or a right to a possession, even if it can be proven that one does not really own that possession in the first place? In order to ensure that depropriation is not experienced as a traumatic violence enabling further attempts at reappropriation of subject, object, property, or commodity, the adoption of certain practices concerning copying would be beneficial. Depropriation, as Simone Weil recognized, has to begin close to home, with our own selves and communities and the objects we use to prop up limited notions of identity. In this regard, the Dalai Lama's current position—his refusal to seek Tibet's total political independence from China—can be viewed as a powerful example of the politics of depropriation: a refusal to buy into the appropriative discourse and politics of modern nationhood, and, at the same time, an effort to safeguard the contingent, historical identities and autonomy of the diversity of peoples named "Tibetan."[36]

Depropriation should usually be consensual (though, as one of my Buddhist teachers said, sometimes you have to snatch madness away from someone if you can). And it should be something around which actual communities form (these I call "folk cultures," for reasons set out earlier), in recognition of the dynamic contingency and impermanence of particular framings of subject and object. Depropriation means learning how to relax—just as Khenpo Tsültrim suggested when I rode the roller coaster at Space Mountain. It means maintaining a relaxed but disciplined attitude toward phenomena—toward phenomena as they arise for us in the situation we find ourselves in. No "copy" labeled as such can ever "be" nondual, but it can be the mark of our yearning for, and part of a practice that leads to a recognition of, "it." Struggling with words, ideas, concepts, copying can only be a misrecognition—yet it is one that brings us closer to a realization of *śūnyatā,* if we pay attention closely enough.

Coda

From the Right to Copy to Practices of Copying

The copy shop in Toronto where I've had coursepacks made for a number of years was busted recently, and the books used to make the coursepacks were confiscated, along with the coursepacks themselves. The store's owner gave me the number of Access Copyright, the organization responsible for the bust. When I called the number and spoke to one of the agents there, I was informed that the copy shop apparently lacked a license to make coursepacks, and that in the future I should frequent copy shops that have licenses. My books were shipped back to me, along with a list of legitimate copy shops, whose owners responded in an understandably cautious and suspicious way when I contacted them to see if they were indeed Access Copyright licensed. After many phone calls, requests for information, and the like, a copy shop near York University reluctantly produced my coursepacks, at a price approximately four times higher than before. And so, in planning my winter courses, I decided to give my copy business to the university bookstore, with its six-week turnaround time, high prices, and extraordinary restrictions on the types and quantities of materials that faculty members are allowed to teach.

This was something of a rude awakening for the author of the preceding pages, which have argued that debates concerning the legal framing of copying within a system of rights and property miss the universal nature of processes of imitation and copying that help constitute the very possibility of being human, inhabiting the world, positing the existence of subject and object, and other core framings of our life situation. Although I am broadly sympathetic to the liberal critique of existing intellectual-property law, embodied in recent works such as Lawrence Lessig's *Free Culture* and James Boyle's *Public Domain,* these critiques accept the capitalist system as it currently stands, and propose modifications of IP law that basically support the expansion of that system and its need to exploit creative labor, entrepreneurship of ideas, and so on. I have argued that if those seeking a "free culture" can posit the freedom of culture only in terms of the existing system, then how free can such a culture really be? Furthermore, the actual practices of copying that are found on peer-to-peer networks—in the promiscuous exchange of ideas, sounds, languages, and bodies in carnivals, dance halls, art events, and other such contemporary spaces—are poorly described by the discourses of entrepreneurship or creative labor. They point toward a freedom that is joyful, and that does not depend on the law. Was I wrong in bracketing the law as a secondary concern when thinking about copying?

My anecdote about coursepacks is hardly unusual—just another minor skirmish in the intellectual-property wars. Yet it is very revealing of the situation regarding copying in North America at the turn of the twenty-first century. The distribution of learning materials within the educational system was considered a special case even before the birth of copyright law in Britain in 1709, with the Statute of Anne. For example, as far back as 1610 there had been regulations stipulating that copies of all published works be deposited in the libraries of the great European universities such as Oxford.[1] In certain countries such as the United States, a fair use exception to copyright has protected scholars who need to make copies of texts for research or study, but education has increasingly fallen within the domain of the marketplace, where such learning materials are today considered private property which requires permission in order that access be granted, and a corresponding fee levied for use. The more restrictive fair dealing exception in Canadian law offers weaker protection to scholars and researchers and in the corporate Canadian university of today, the interpretation and enactment of intellectual-property laws, which would previously have been carried out by law agencies directly acting on behalf of the nation-state, are today increasingly enacted by private organizations such as Access Copyright that are funded by and act as proxies for industry. Indeed it is striking that in both newspaper coverage of coursepack busts and Access Copyright's own press releases, which routinely describe unlicensed copying of coursepacks as "robbery" or "piracy", there is rarely if ever even mention of the existence of a fair dealing and/or fair use exception.[2]

Access Copyright, formerly known as Cancopy, has its origin in a provision of the Canadian law that allows for "collectives" to administer permissions and fees on behalf of a wide range of copyright holders. To quote the organization's website: "Since 1988, Access Copyright has been meeting the needs of businesses, educators, governments and other organizations across Canada with our innovative copyright licencing solutions. Our licences give content users immediate, legal access to the copyright protected materials they need to copy from to get their jobs done, while ensuring that creators and publishers are fairly compensated when their works are copied."[3] This apparently straightforward and reasonable statement of "what we do" condenses a good deal of the ideology of intellectual property under late capitalism—a discourse of meeting needs, providing solutions, allowing access, getting jobs done, and of course compensating all interested parties fairly. The term "fair dealing" is itself a transposition from the British copyright law of 1911 to the original Canadian Copyright Act of 1921; and the word "fair" has a long and complicated history, moving between its traditional senses of "beautiful" and "virtuous" to a more particular legal/economic meaning in the eighteenth century, when the term "fair dealing" was first used.[4] The word participates in the rhetoric of impartiality that supported British imperialism, as well as capitalist ethics: after the inaugural act of violence with which one imposes a system, one seeks only fair play—i.e., behavior that accepts the newly imposed norms. According to the discourse that this word is a part of, getting something without paying for it is "unfair," and the idea that people of limited means (students, for example, or users of public libraries) have as much right to access the archive of publicly disseminated works as those who are rich is also "unfair." Conversely, we might offer an alternative definition of "fair" in this context by saying that a progressive and just society allows the free circulation of materials required for the education of its citizens, both in libraries and in the classroom, and that such circulation, in order to be "fair," should not be subject to permissions, royalties, seizure, or arbitrary limits on the number of chapters of a book that can be copied. The fact that copy shops offering reprographic services make a profit in facilitating research and study should not distract us from the real issue here: that the "fair compensation" which Access Copyright pursues puts the private profit of individuals and corporations before the needs of educators, students, and society as a whole, and is in fact unfair.

Having said this, I know that not all students will buy their coursepacks from the university bookstore. Some will make their own photocopies, or scan the coursepacks and distribute them as PDFs. Who knows—maybe they'll write the texts out by hand. Certainly, some students will share a single copy. Others, deterred by the high price, won't bother to buy or read the texts at all. Still others will download PDFs of the course texts they have found by Googling, or they'll use Google Books to "preview" the texts, or (as at least one of my students has done) they'll skip the readings and find synopses on Wikipedia or some other website. It is even possible that one or two might visit the university library and find the texts there. The proliferation of copies proceeds "asymmetrically," regardless of the wishes of regulators or, for that matter, legal producers. I now get official coursepacks made, but students still find ways around buying them.

Yet, when it comes down to it, in this situation I obeyed the law. We must ask: Can there be such a thing as free copying without a profound confrontation with the law? Isn't this another instance of the battle in which the poststructuralist Deleuzians and Derrideans square off against structuralists such as Žižek and Badiou? The former affirm the dissolution of hegemonic structures within universal rhizomatic processes of playful assemblage and disassemblage, which can be held only momentarily within the illusory framings of discourse, ideology, law, and structure. The latter, in contrast, insist on the reality of the symbolic structures of Law, and the necessity of recognizing and confronting such structures in order to enact changes that cannot easily be appropriated back into an otherwise unchallenged logic of Capital.

Copying is always already a crucial aspect of our ability to articulate ourselves and our world. Language functions mimetically, and therefore discourse, ideology, self-expression, community are also mimetic. The same is true for the university. As Kate Eichhorn has argued in her study of copy shops around the University of Toronto, historically universities have always relied on those who provide copying services (this was true even in medieval times), whether legal or not.[5] Put simply: there is no university without copying, since the university's mandate is itself disseminative mimesis. Yet scholars have framed the universality of copying in quite specific ways within modernity, usually by obscuring the active constitutive presence of mimesis in discourses, identities, structures, and institutions, in order to naturalize them. At the same time, modernity offers the spectacle of a series of abjected, overdetermined, scapegoated mimetic threats that always appear to come from the outside, from the margins, threatening to contaminate and infect that pure, copy-free world of originals that we are told we inhabit. Such threats include: the foreigner as the nonhuman, inauthentic usurper who pretends to be like us; the feminine as the hysterical, irrational, duplicitous, seductive power of the false; the drug as inducer of a simulacrum of pleasure and happiness that leads to ruin; the counterfeiter, the pirate, the mafia as criminals who infiltrate legitimate economies with illegitimate fake products.

I have argued that we cannot actually live in a world without mimesis. For Locke and for Marx, appropriation is constitutive of being-in-the-world through labor or sensuous activity; for Hegel, property, ownership of self, is the basis of society.[6] Both appropriation in general, and ownership as a particular form of appropriation, are mimetic in that they bestow a particular name on something—a name that identifies and frames it. The named, labeled, identified form (including that of the subject, i.e., ourselves) is always already a copy. The various trajectories of twentieth-century philosophy and theory, from Bataille, Heidegger, Beauvoir, and the Frankfurt School through Foucault, Derrida, Butler, and Spivak, have taken apart the residual overdeterminations of mimesis that were already fully articulated by Plato. Their texts contain powerful critiques of intellectual property and the struggle to articulate a different basis for understanding identity, action, and community—but with a few exceptions, these aspects of critical theory have not been fully developed.[7] The core issue that they address is that of a universal flux, a "chaosmos"—in Buddhist terms, universal impermanence and interdependence. Mimesis in the Platonic sense articulates the desire to fix this flux permanently, but it could equally be understood as the radiant, ever-shifting flux "itself" in its infinite transformations and appearances.

The proliferation of copies that is occurring in contemporary IP struggles is multivalent. Of course, one has little urge to condone the fact that mafias in various countries make MP3s available on the cheap. On the other hand, the enforcers of copyright, whether overzealous publishers, estates, or law-enforcement organizations, are often equally difficult to admire. Nevertheless, this "war" over intellectual property, the endless examples of conflict over the right to copy that fill the press and academic journals and conferences devoted to the topic, even the example of my own struggles with which I began this coda—aren't they a distraction from the omnipresence of mimesis? By limiting the analysis of copying to a very circumscribed set of situations, this debate risks obscuring something much more troubling and powerful—which nevertheless asserts itself in every controversy concerning intellectual property. Mimesis is the "accursed share" that Bataille wrote of: the force or quality of the universe that exceeds us in every way, yet impels us to act, to respond, to frame.[8] Or does it? It is possible to think beyond or through the frameworks of appropriation that support concepts of property, intellectual or otherwise, and to move toward a depropriated subject and object. But it is challenging, too. Modernity is built around certain structurings of mimesis; to change our world today necessarily means to go beyond such structurings. Indeed, I argue that various traditional cultures—notably, for my own work, those associated with Buddhism—are built around an ethics and practice that, while often falling sadly short in existing Buddhist societies, nevertheless articulate a vision of a universe and collectivity that actively engages and works with mimesis while abandoning all notions of property at their illusory roots.

It sometimes seems that we live impossibly far from any such utopian vision of equality, justice, and community, and therefore should limit any discussion to modifications of the existing legal structure. But again, I must insist that the legal domains in which copying is framed are themselves mimetic structures. Law as institution, as intervention, as structure, exists so that society can place limits on ubiquitous, omnipresent mimetic transformation. Copying occurs inside those domains, outside their precincts, and in the construction of boundaries, definitions, which produce an inside and an outside. For example, Access Copyright's "Captain Copyright" campaign against "illegal" copying used text that had been copied from public-domain materials without attribution.[9] How the boundaries that establish law are constructed—what counts as inside or outside, legitimate or illegitimate, original or copy—matters. Yet the persistence of copying points to something else. What a copy "is" depends on whether and how property and rights, which have particular histories, are defined, and on the community that defines or does not define them. What the Internet offers us is not so much new forms of economy, production, and exchange (although the open-source movement has certainly made efforts in those directions), but the opportunity to render visible once more the instability of all the terms and structures which hold together existing intellectual-property regimes, and to point to the madness of modern, capitalist framings of property. In this way, contemporary struggles over IP rights link up with a broad range of modern critiques of property, intellectual or otherwise, from the work of the theorists listed above; with the artistic avant-gardes; with folk cultures, traditional, subcultural, and otherwise; with the struggles of explicitly political groups, ranging from communists to followers of Gandhi to anarchists; with the feminist critique of identity and objecthood.

I believe it is a mistake to assume, as most liberal critiques of existing IP law do, that intellectual property and property *qua* property, material or otherwise, should be treated differently. While there are differences between physical and intellectual property, the problem, at least at the level of contemporary legal-political discourse, is property, intellectual or otherwise, and the systems and structures that govern property. Indeed, what if all property were actually "intellectual property"—in other words, a conceptual fabrication or work of imagination, rather than a fact? Within the discourse of property and rights, "fair use" and "the public domain" are crippled concepts unless they include, for example, the right to cross national borders (fair use of land), or access to food, hospitals, medicine, and education (all of which have been, to different degrees, parts of public domains at some time or another). One possible and provisional answer to many of the problems that plague humanity today, particularly those predicated on scarcity, is simply to *make more copies* and distribute them freely—as in the story of Jesus and the feeding of the five thousand. And from a more fundamental perspective, this would already be a gesture in the direction of depropriation.

But is the core issue concerning intellectual property really that of the "right to copy"? It will be argued that if we give up talking about rights, and about the structures that guarantee them, we are left in Hobbes's state of nature, the kind of chaos in which the worst predators, those who are able to most aggressively appropriate, would dominate. Thus, Hegel spoke of right as fundamental to the constitution of a person and a progressive society.[10] But a human being is not just a "bundle of rights," to use the legal expression. A bundle is already a montage, a little package of chaos; and that montage consists of a cloud of transforming repetitions, whose direction, insofar as it is conscious, is a matter of practice. A full analysis of right as it relates to imitation is beyond the scope of this coda. Nevertheless, it is possible to think of copying outside the realm of right and ownership if we conceive of copying as a practice, or rather a multitude of practices. More important, not only is it possible to think this way, but historically many communities actually have done so. That practice and right are different is indicated by the many stories of great folk artists and masters—musicians, yogis, warriors, lovers—who stole their knowledge from official sources in order to teach it to others. Practice is highly mimetic, is eminently transportable, and belongs to no one, despite all dogma to the contrary. It is a matter of value and competence, rather than right. One does not need to own in order to practice; if anything, a practice owns us, reshapes and reconfigures us, and inserts us in a dynamic collectivity. Practice has its own ethics—and this ethics is worked out in the configuration of practice itself, and in relation to other practices and practitioners.

The concept of practice is oddly underdeveloped within the Western philosophical tradition, despite being central to the major modern ruptures of that tradition which form the corpus of critical theory and the artistic avant-gardes. It is also copying as practice that sustains folk communities, ranging from traditional societies through punk and hip-hop to whatever is today labeled "subculture." A particular practice of copying likewise sustains various forms of capitalist economy. The impasse of the avant-gardes can be seen in the Situationist call for the creation of "new situations," which has produced a vast accumulation of gestures now contained in the huge bubble that is the gallery and museum system. Subcultures, on the other hand, have developed powerful practices—notably in the fields of music, style, and community—which burst forth as "temporary autonomous zones," to use Hakim Bey's phrase, but are either appropriated into the existing hegemonic mainstream or safely bracketed at the margins of society.[11] While I generally agree with Hardt and Negri's formulation of the multitude against empire, articulating a new vision of a common wealth, what is strikingly absent in their work is any sense that revolutionary communities have any positive content other than "resistance."[12]

We need a reinvigorated, critical concept of practice in cultural and political theory. More important, we need to recognize the way that practices of copying are continually being negotiated and refined by marginal communities today—defensively, in response to a global political and economic system that exploits them, but also autonomously, joyfully, as ends in themselves. Yes, the factory worker in Shenzhen, the cumbia musician in Buenos Aires, the rapper in Angola, the student at the copy shop in Toronto need to understand their commonality and the possibility of collective action. But the struggle to affirm the most valuable, most enriching forms of practice can lead us beyond the modern formulations of right, property, ownership, and copyright. This struggle can and does begin with the most trivial everyday incidents—the price of an academic course-pack, the right to use a Disney character in a story published on the Internet, the availability of cheap, fake designer bags. In each case, it is the absurd overdetermination and enforcement of an unjust law that illuminates for ordinary citizens the reality of the existing regime. Conversely, every unjust legal intervention draws attention to the trivial but inexorable freedom that underlies our ability to act as individuals and communities in everyday life, and invites us to investigate it, familiarize ourselves with it, and realize it, as individuals and communities, in practice.

Notes

Acknowledgments

Index

Notes

Introduction

1. Pierre Recht, *Le Droit d'auteur, une nouvelle forme de propriété* (Paris: Librairie Générale de Droit et de Jurisprudence, 1969), pp. 27–47; Michael Newcity, *Copyright Law in the Soviet Union* (New York: Praeger, 1978), p. 19.

1. What Is a Copy?

1. See Guy Trebay, "This Is Not a Sidewalk Bag," *New York Times,* April 6, 2008, www.nytimes.com/2008/04/06/fashion/06brooklyn.html; and Susan Scafidi, "Louis Vuitton's Fake Fighting Parody," *Counterfeit Chic,* April 3, 2008, www.counterfeitchic.com/2008/04/post_13.php (both accessed October 24, 2008), for accounts of the opening night. I visited the show after the opening night.

2. Michelle Bery, "Imitating a Brand with Louis Vuitton Replica Handbags," n.d., ezinearticles.com/?Imitating-A-Brand-With-Louis-Vuitton-Replica-Handbags&id=599644 (accessed October 24, 2008); Melissa Hancock, "Fighting the Fakes," January 7, 2007, www.arabianbusiness.com/5776?tmpl=component&page= (accessed October 24, 2008).

3. See Wikipedia entry for Louis Vuitton.

4. Hsiao-hung Chang, "Fake Logos, Fake Theory, Fake Globalization," *Inter-Asia Cultural Studies,* 5, no. 3 (2004), pp. 222–236.

5. "Fakewear," on Mind What You Wear site, www.mindwhatyouwear.com/fake1.html (accessed October 24, 2008).

6. Andrew Mertha, *The Politics of Piracy: Intellectual Property in Contemporary China* (Ithaca, N.Y.: Cornell University Press, 2005), p. 165.

7. Tim Phillips, *Knockoff: The Deadly Trade in Counterfeit Goods* (London: Kogan Page, 2005), p. 34.

8. Ibid., pp. 25–27.

9. Murakami canvases that feature Louis Vuitton logos include *The World of Sphere* (2003), *Eye Love Superflat* (2003 and 2004 versions), and the "monogramouflage" series exhibited in the boutique at the Brooklyn Museum.

10. On Wilson: Scott Rothkopf, "In the Bag," *Artforum*, September 2003. On Park: www.counterfeitchic.com/2007/01/is_it_or_isnt_it_2.php. Architect Rem Koolhaas also produced a series of fake ads for Prada involving counterfeit bags; the ads appeared in 2006 in his book/magazine *Content.*

11. See Sarah McCartney, *The Fake Factor: Why We Love Brands but Buy Fakes* (London: Cyan, 2005).

12. Lynn Yaeger, "Purse Snatching: What's a Girl Got To Do To Get a Fake Louis Vuitton Around Here?" *Village Voice,* July 17, 2007, villagevoice.com/2007-07-17/nyc-life/purse-snatching/ (accessed October 24, 2008).

13. BasicReplica website, www.basicreplica.com/ (accessed October 24, 2008). In March 2009 the owners of Basicreplica.com were ordered by a Florida court to stop selling replicas, and after the owners ignored this order, the domain name was transferred to the plaintiffs in the case, Chanel and LVMH, who posted the court order on the website (website accessed again March 12, 2010).

14. Göran Sörbom, *Mimesis and Art: Studies in the Origin and Early Development of an Aesthetic Vocabulary* (Stockholm: Svenska Bokoförlaget, 1966), p. 100.

15. Martin Heidegger, *Nietzsche,* vol. 1 (San Francisco: Harper and Row, 1991), p. 173. Subsequent references to this work appear in parentheses in the text.

16. Www.basicreplica.com/faq.htm#1 (accessed October 26, 2008).

17. David R. Koepsell, *The Ontology of Cyberspace: Philosophy, Law, and the Future of Intellectual Property* (Chicago: Open Court, 2000), proposes to answer this question, but the book's understanding of ontology is rather limited and does not go beyond thinking of IP in terms of categories.

18. "Twice Bitten: Louis Vuitton v. Haute Diggity Dog," *Counterfeit*

Chic, November 2007, www.counterfeitchic.com/2007/11/twice_bitten_louis_vuit ton_v_h.php (accessed May 22, 2009).

19. On the complexity of Aristotle's conceptions of essence and idea, see Charlotte Witt, *Substance and Essence in Aristotle* (Ithaca, N.Y.: Cornell University Press, 1989); and the reconstruction of Aristotle's lost work *On Ideas* in Gail Fine, *On Ideas: Aristotle's Criticism of Plato's Theory of Forms* (New York: Oxford University Press, 1993).

20. For a thorough chronological account of the philosophical concept of mimesis, see Gunter Gebauer and Christolph Wulf, *Mimesis: Culture, Art, Society*, trans. Don Reneau (Berkeley: University of California Press, 1995). For an account of the history of mimesis in Western literature, see Eric Auerbach, *Mimesis: The Representation of Reality in Western Literature*, trans. Willard R. Trask (Princeton: Princeton University Press, 1953). For a decidedly nonchronological account of the mimetic diaspora, see Hillel Schwartz's epic work *The Culture of the Copy: Striking Likenesses, Unreasonable Facsimiles* (New York: Zone, 1996).

21. See John Milbank and Catherine Pickstock, *Truth in Aquinas* (London: Routledge, 1999), p. 87.

22. Badiou's critique of Deleuze is astute in pointing out that for all his avowed anti-Platonism, Deleuze ultimately proposes "a Platonism of the virtual"; see Alain Badiou, *Deleuze: The Clamor of Being*, trans. Louise Burchill (Minneapolis: University of Minnesota Press, 2000), p. 45. Yet Badiou's attempts at recuperating Plato through set theory are not entirely successful either, amounting to a Platonism of the event, with mathematical ontology serving as "idea"; see Alain Badiou, *Being and Event* (New York: Continuum, 2005).

23. William P. Alford, *To Steal a Book Is an Elegant Offense: Intellectual Property Law in Chinese Civilization* (Stanford, Calif.: Stanford University Press, 1995); Siva Vaidhyanathan, *Copyrights and Copywrongs: The Rise of Intellectual Property and How It Threatens Creativity* (New York: New York University Press, 2001).

24. Timothy Morton, "Hegel on Buddhism," *Romantic Circles*, February 2007, special issue on Romanticism and Buddhism, ed. Mark Lussier, at www.rc.umd.edu/praxis/buddhism/morton/morton.html (accessed October 20, 2009). See also Lawrence Sutin, *All Is Change: The Two-Thousand-Year Journey of Buddhism to the West* (New York: Little, Brown, 2006), pp. 144–170.

25. See Keiji Nishitani, *Religion and Nothingness*, trans. Jan Van Bragt

(Berkeley: University of California, 1982); Sri Aurobindo, "Art," in Aurobindo, *Complete Works,* vol. 1 (Pondicherry: Sri Aurobindo Ashram, 1970–1975), pp. 538–541; and "Indian Art," in *The Foundations of Indian Culture,* 3rd ed. (Pondicherry: Sri Aurobindo Ashram, 1971), pp. 248–249.

26. Thomas McEvilley, *The Shape of Ancient Thought: Comparative Studies in Greek and Ancient Thought* (New York: Allworth, 2002).

27. The core Mahayana teachings (including the famous Heart Sutra) are found in the *Prajñāpāramitā Sutras.* I first encountered them through Lex Hixon's marvelous retranslation of older translations, *Mother of the Buddhas: Meditation on the Prajnaparamita Sutra* (Wheaton, Ill.: Quest, 1993). The sutras are the teachings of the Buddha. There is a vast philosophical literature, both classical and modern, reflecting upon emptiness. My own understanding comes from the oral and written teachings of Khenpo Tsültrim Gyamtso Rinpoche, notably in his books *The Sun of Wisdom: Teachings on the Noble Nagarjuna's Fundamental Wisdom of the Middle Way,* trans. Ari Goldfield (Boston: Shambhala, 2003), and *Progressive Stages of Meditation on Emptiness* (Auckland, N.Z.: Zhyisil Chokyi Ghatsal, 2001).

28. See Roger Pol-Droit, *The Cult of Nothingness: The Philosophers and the Buddha* (Chapel Hill: University of North Carolina Press, 2003), on the history of Western philosophical misinterpretations of emptiness.

29. Youxuan Wang, *Buddhism and Deconstruction: Towards a Comparative Semiotics* (Richmond, Surrey: Curzon, 2001); Ryūichi Abé, *The Weaving of Mantra: Kūkai and the Construction of Esoteric Buddhist Discourse* (New York: Columbia University Press, 1999).

30. Stéphanie Bonvicini, *Louis Vuitton: Une Saga française* (Paris: Fayard, 2004), pp. 224–225.

31. Both quotes are taken from Jacques Derrida, "Differance," in Derrida, *Speech and Phenomena,* trans. David B. Allison (Evanston, Ill.: Northwestern University Press, 1973), p. 148.

32. Alain Badiou, *Ethics: An Essay on the Understanding of Evil,* trans. Peter Hallward (London: Verso, 2001), p. 25.

33. Walter Benjamin, "Doctrine of the Similar," in Benjamin, *Selected Writings,* vol. 2, ed. Michael W. Jennings, Howard Eiland, and Gary Smith (Cambridge, Mass.: Harvard University Press, 1999), pp. 694–698; Benjamin, "On the Mimetic Faculty," ibid., pp. 720–722.

34. Walter Benjamin, "Antitheses Concerning Word and Name," ibid., p. 717.

35. Walter Benjamin, *On Hashish,* trans. Howard Eiland and others (Cambridge, Mass.: Harvard University Press, 2006), p. 53 (protocol for September 29, 1928) and p. 123 ("Hashish in Marseilles").

36. See Morton, "Hegel on Buddhism."

37. Examples, in the history of Tibetan Buddhisms, include the Two Truths debate, the distinctions between Svātantrika and Prāsaṅgika Madhyamaka and between Shentong and Rangtong Madhyamaka, and the suppression of the Jonangpa sect by the Gelukpas. See *The Svātantrika-Prāsaṅgika Distinction: What Difference Does a Difference Make?* ed. Georges B. J. Dreyfus and Sara L. McClintock (Somerville, Mass.: Wisdom, 2003); and Khenpo Tsültrim Gyamtso Rinpoche, *Progressive Meditation on the Stages of Emptiness* (Auckland, N.Z.: Zhyisil Choky Ghatsal, 2001), on the distinctions between different Buddhist schools regarding emptiness.

38. See Robert Magliola, *Derrida on the Mend* (West Lafayette, Ind.: Purdue University Press, 1984), pp. 127–129, on Derrida and suchness, or *tathātā;* Wang, *Buddhism and Deconstruction,* on "the Same"; and David Loy, *Nonduality: A Study in Comparative Philosophy* (Atlantic Highlands, N.J.: Humanities Press, 1999), pp. 249–259, on Derrida's incomplete deconstruction of dualisms.

39. Wang, *Buddhism and Deconstruction,* pp. 125–126. Jay Garfield and Graham Priest, "Nāgārjuna and the Limits of Thought," in Garfield, *Empty Words: Buddhist Philosophy and Cross-Cultural Interpretation* (Oxford: Oxford University Press, 2002), p. 105.

40. A number of theorists, including Slavoj Žižek in *The Parallax View* (Cambridge, Mass.: MIT Press, 2006), have recently argued that it is the inbetween or the relational that constitutes a kind of functioning ground or essence—but the reification of the inbetween is as problematic as the notion of essence. There is no inbetween. What there is, is the interdependent constellation of the mimetic at the relative level, and concept-free nonduality at the absolute level.

41. Robert Magliola, "Afterword," in Jin Park, ed., *Buddhisms and Deconstructions* (Lanham, Md.: Rowman and Littlefield, 2006), p. 243.

42. Michael Taussig, *Mimesis and Alterity: A Particular History of the Senses* (New York: Routledge, 1993), p. 52. Subsequent references to this work appear in parentheses in the text.

43. Chang Yen Yuan, in *Early Chinese Texts on Painting,* ed. Susan Bush and Hsio-yen Shih (Cambridge, Mass.: Harvard University Press, 1985), p. 54.

44. Giordano Bruno, "A General Account of Bonding," in *Cause, Principle, and Unity,* trans. Richard Blackwell (Cambridge: Cambridge University Press, 1998).

45. See the chapter "Mimetic Desire" in René Girard, *Things Hidden since the Foundation of the World,* trans. Stephen Bann and Michael Metteer (Stanford, Calif.: Stanford University Press, 1987), pp. 283–298.

46. Jacques Derrida, "Economimesis," *Diacritics,* 11 (1981), pp. 3–25.

47. Martha Woodmansee, *The Author, Art, and the Market: Rereading the History of Aesthetics* (New York: Columbia University Press, 1994).

2. Copia, or, The Abundant Style

1. *Oxford English Dictionary Online,* 2nd ed. (1989), entry for "copy, *n.*" (accessed October 25, 2008).

2. Ovid, *Metamorphoses,* trans. Mary Innes (London: Penguin, 1955), p. 205.

3. Arkady and Boris Strugatsky, *Roadside Picnic / Tale of the Troika* (New York: Pocket Books, 1977), p. 151.

4. *A Latin Dictionary Founded on Andrews' Edition of Freund's Latin Dictionary, Rev., Enl., and in Great Part Rewritten by Charlton T. Lewis and Charles Short* (Oxford: Clarendon, 1958), pp. 466–467.

5. *Oxford Latin Dictionary* (Oxford: Clarendon, 1968), vol. 1, pp. 442–443.

6. On "copia" and "Ops," see *Brill's New Pauly: Encyclopaedia of the Ancient World,* ed. Hubert Cancik and Helmut Schneider (Boston: Brill, 2002–), vol. 3, p. 765, and vol. 10, p. 172. For a more extensive discussion of Ops, see Pierre Pouthier, *Ops et la conception divine de l'abondance dans la religion romaine jusqu'à la mort d'Auguste* (Paris: Diffusion de Boccard, 1981); and Georges Dumezil, *Idées romaines* (Paris: Gallimard, 1969), pp. 289–304. The exact nature of the relation between Ops and Consus, according to these latter authors, is a matter of debate, but we know that Ops was worshiped as Ops Consiua in a shrine in the Regia in Rome. On Ops and copia, see Pouthier, *Ops,* p. 23: "The root has a rich Latin descendance in the form of derivations of 'opus,' indicating 'the product of activity,' while the word 'ops' itself is the source of numerous derivatives and compound forms, among which is 'copia,' with the parallel meaning of 'abundance,' 'resource,' 'securing.'" (Unless otherwise stated, all translations are my own.)

7. See Margarete Bieber, *Ancient Copies: Contributions to the History of*

Greek and Roman Art (New York: Columbia University Press, 1977); and Ellen Perry, *The Aesthetics of Emulation in the Visual Arts of Ancient Rome* (Cambridge: Cambridge University Press, 2005).

8. Jacques Derrida, "Economimesis," *Diacritics*, 11 (1981).

9. Erasmus, *On Copia of Words and Ideas*, trans. Donald B. King and H. David Rix (Milwaukee: Marquette University Press, 1963), p. 9.

10. See Book 10 of Quintilian's *Institutio Oratorio.*

11. *Oxford English Dictionary Online.*

12. See John Feather, "From Rights in Copies to Copyright: The Recognition of Authors' Rights in English Law and Practice in the Sixteenth and Seventeenth Centuries," in *The Construction of Authorship: Textual Appropriation in Law and Literature*, ed. Martha Woodmansee and Peter Jaszi (Durham: Duke University Press, 1994), p. 192.

13. See Ronan Deazley, *On the Origin of the Right to Copy: Charting the Movement of Copyright Law in Eighteenth Century Britain, 1695–1775* (Oxford: Hart, 2004); and Martha Woodmansee, *The Author, Art, and the Market: Rereading the History of Aesthetics* (New York: Columbia University Press, 1994).

14. Edward Young, *Conjectures on Original Composition*, cited in David Goldstein, entry on "Originality," in *The Princeton Encyclopedia of Poetry and Poetics*, 4th ed., in press.

15. Rosalind Krauss, "The Originality of the Avant-Garde," in *The Originality of the Avant-Garde and Other Modernist Myths* (Cambridge, Mass.: MIT Press, 1985), p. 167.

16. John Cage, *Silence: Lectures and Writings by John Cage* (Cambridge, Mass.: MIT Press, 1966); William S. Burroughs, *The Job: Interviews with William S. Burroughs by Daniel Odier* (New York: Grove, 1974); Andy Warhol, *The Philosophy of Andy Warhol: From A to B and Back Again* (New York: Harcourt Brace Jovanovich, 1975).

17. Terence Cave, *The Cornucopian Text: Problems of Writing in the French Renaissance* (New York: Oxford University Press, 1979).

18. Mikhail Bakhtin, *Rabelais and His World*, trans. Helene Iswolsky (Cambridge, Mass.: MIT Press, 1968).

19. Gayatri Spivak, *Death of a Discipline* (New York: Columbia University Press, 2003), p. 16.

20. Thurston Moore, ed., *Mix Tape: The Art of Cassette Culture* (New York: Universe, 2004), p. 28.

21. Walter Benjamin, *On Hashish,* trans. Howard Eiland and others (Cambridge, Mass.: Harvard University Press, 2006), pp. 58–60, 81–82.

22. Ananda Coomaraswamy, *The Dance of Śiva: Essays on Indian Art and Culture* (Mineola, N.Y.: Dover, 1985; orig. pub. 1924), pp. 30–45.

23. Thanks to Erik Davis for pointing out the sutra as a key figuration of multiplicity. See Davis, *Techgnosis: Myth, Magic and Mysticism in the Age of Information* (London: Serpent's Tail, 1999), p. 319.

24. *The Flower Ornament Scripture: A Translation of the Avatamsaka Sutra,* trans. Thomas Cleary (Boulder, Colo.: Shambhala, 1984), vol. 1, p. 207.

25. Lothar Ledderose, *Ten Thousand Things: Module and Mass Production in Chinese Art* (Princeton, N.J.: Princeton University Press, 2000), p. 151.

26. John Kieschnick, *The Impact of Buddhism on Chinese Material Culture* (Princeton: Princeton University Press, 2003), pp. 164–185.

27. On the history of making multiples of objects as a way of gaining merit in China, see ibid., pp. 157–164.

28. As Ledderose writes in *Ten Thousand Things,* p. 152: "To Buddhists, the copying of texts was a means toward acquiring merit. The more copies, the more merit, the ultimate merit being liberation from the circle of rebirth. Printing made it easy to produce multiple copies—it was fast, cheap, and more convenient than copying texts by hand. Moreover, it avoided all the mistakes a copyist might make, and that was of special importance in the case of sacred invocations. It was not deemed obligatory that every copy of a text be recited. Its mere existence conveyed bliss to donor and owner alike. This is one reason why Buddhist printers often strove for huge editions."

29. Cited and translated by Kieschnick, *The Impact of Buddhism,* pp. 168–169.

30. Ernst F. Schumacher, "Buddhist Economics," in Schumacher, *Small Is Beautiful: Economics As If People Mattered* (New York: Harper and Row, 1973).

31. For Deleuze, the event is the moment of instantiation of the actual within, and as a part of the field of the virtual, while for Badiou, the event is the emergence or eruption of a pure outside that sets off truth procedures and the production of multiplicities which track or articulate it. My own position is closer to that of Deleuze, although I find the terms "actual" and "virtual" misleading. See Alain Badiou, *Logics of Worlds: Being and Event II,* trans. Alberto Toscano (London: Continuum, 2009), pp. 381–389; Gilles Deleuze, *The Logic of Sense,* trans. Mark Lester and Charles Stivale (New York: Colum-

bia University Press, 1990), pp. 148–153; and Gilles Deleuze and Claire Parnet, "The Actual and the Virtual," in Deleuze and Parnet, *Dialogues II* (London: Continuum, 2002), pp. 112–115.

32. *Rhythm Shower* and *Version Like Rain* are both titles of LPs by Jamaican producer Lee Perry. The latter recording is a considerable collection of tracks that are all versions of three particular rhythms.

33. Houston Baker, "Hybridity, the Rap Race, and Pedagogy for the 1990s," in *Technoculture,* ed. Constance Penley and Andrew Ross (Minneapolis: University of Minnesota Press, 1991), p. 200.

34. Paul D. Miller points to this mathematical heritage of hip-hop in his book *Rhythm Science* (Cambridge, Mass.: Mediawork / MIT Press, 2004), and the possibility of "making a world" out of the montage and combination of two records.

35. Jeff Chang, *Can't Stop, Won't Stop: A History of the Hip-hop Generation* (New York: St. Martin's, 2005), p. 43.

36. Michael Warner, *Publics and Counterpublics* (Cambridge, Mass.: Zone Books, 2002).

37. On this topic, see Virginia Postrel, *The Substance of Style: How the Rise of Aesthetic Value Is Remaking Commerce, Culture, and Consciousness* (New York: HarperCollins, 2003).

38. Paul Gilroy, *The Black Atlantic: Modernity and Double Consciousness* (Cambridge, Mass.: Harvard University Press, 1993); Kodwo Eshun, *More Brilliant than the Sun: Adventures in Sonic Fiction* (London: Quartet, 1998).

39. Greil Marcus, *Invisible Republic: Bob Dylan's Basement Tapes* (New York: Holt, 1997).

40. Ernesto Laclau, *On Populist Reason* (London: Verso, 2005), p. 63.

41. See Marcus Boon, "Sublime Frequencies Ethnopsychedelic Montages," *Electronic Book Review,* Musicsoundnoise issue, December 2006, www.electronicbookreview.com/thread/musicsoundnoise/ethnopsyche (accessed October 23, 2009); and idem, "Carnival Folklore Resurrection in the Age of Globalization," www.marcusboon.com (accessed October 23, 2009).

3. Copying as Transformation

1. *The Complete Works of Chuang Tzu,* trans. Burton Watson (New York: Columbia University Press, 1968), p. 49. The last lines of A. C. Graham's widely respected translation are somewhat different: "Between Chou and the

butterfly there was necessarily a dividing; just this is what is meant by the transformation of things." But this translation somewhat loses the irony and paradox of the story, since the necessity of the dividing is precisely what is being questioned. *Chuang-Tzu: The Inner Chapters,* trans. A. C. Graham (London: HarperCollins, 1991), p. 61.

2. Jacques Lacan, *The Four Fundamental Concepts of Psycho-analysis,* trans. Alan Sheridan (London: Hogarth, 1977), p. 76; Slavoj Žižek, *The Ticklish Subject* (London: Verso, 2000), p. 398.

3. In what follows, I've found Marina Warner's *Fantastic Metamorphoses, Other Worlds: Ways of Telling the Self* (New York: Oxford University Press, 2002) useful in thinking through transformation. Late in the process of writing this book, I encountered George Kubler's admirable meditation on art and temporality, *The Shape of Time: Remarks on the History of Things* (New Haven: Yale University Press, 1962), which also resonates in a variety of important ways with my work here.

4. Gilles Deleuze, *Difference and Repetition,* trans. Paul Patton (London: Continuum, 2004), p. xvi. But the insight is a repetition of Gabriel Tarde's argument in *The Laws of Imitation,* trans. Elsie Clews Parson (New York: Holt, 1903), p. 7: "Repetition exists, then, for the sake of variation."

5. Tarde (like Girard and Taussig) resorts to talk of "contagion" in explaining a mimesis that is invisible in its work and beyond conscious processes of representation. But without affirming emptiness, as I have repeatedly done here, it is unclear how contagion as a process could actually function.

6. Georges Dreyfus, *Recognizing Reality: Dharmakīrti's Philosophy and Its Tibetan Interpretations* (Albany: State University of New York Press, 1997).

7. See F. E. J. Valpy, *An Etymological Dictionary of the Latin Language* (Boston: Adamant Media, Elibron Classics, 2005; orig. pub. London, 1838), p. 101, 6.26.

8. François Jacob, *The Logic of Life: A History of Heredity,* trans. Betty E. Spillmann (New York: Pantheon, 1974), pp. 1 and 9. When he introduces the concept of variety through genetic variation and mutation, however, Jacob does speak of "recopying" (p. 5)—a strange doubling of the concept of copying!

9. Philippe Lacoue-Labarthe, *Typography: Mimesis, Philosophy, Politics,* trans. Christopher Fynsk and others (Cambridge, Mass.: Harvard University

Press, 1989), p. 115. Subsequent references to this work appear in parentheses in the text.

10. Luce Irigaray, *An Ethics of Sexual Difference,* trans. Carolyn Burke and Gillian C. Gill (Ithaca, N.Y.: Cornell University Press, 1993), p. 98.

11. One thinks not only of Malkovich's career as an actor, but also, in the film, of the scheme by Lester, the CEO of LesterFile Corp., for attaining immortality through inhabiting "vessels" such as Malkovich.

12. See Giti Thadani, "The Politics of Identities and Languages: Lesbian Desire in Ancient and Modern Indias," in *Female Desires,* ed. Evelyn Blackwood and Saskia Wieringa (New York: Columbia University Press, 1999).

13. Julia Kristeva makes a similar observation in her essay "Motherhood According to Giovanni Bellini," when, asking how to "verbalize this prelinguistic, unrepresentable memory" that would be maternal sameness, "Heraclitus' flux, Epicurus' atoms, the whirling dust of cabalic, Arab, and Indian mystics, and the stippled drawings of psychedelics—all seem better metaphors than the theories of Being, the logos, and its laws"; see *The Portable Kristeva,* ed. Kelly Oliver (New York: Columbia University Press, 2002), p. 305. Furthermore, her term "chora," introduced in *Revolution in Poetic Language,* attempts to designate "maternal sameness" or "nonsensuous similarity" as a prediscursive flux; but, as she herself notes, the question remains as to how to attribute structure to that which is beyond concepts (ibid., p. 35). The origins of "khora" in Plato's *Timaeus* are also explored by Derrida in "Khora," in Derrida, *On the Name,* trans. David Wood, John P. Leavey Jr., and Ian McLeod (Stanford, Calif.: Stanford University Press, 1995), pp. 92–128.

14. Elias Canetti, *Crowds and Power,* trans. Carol Stewart (New York: Farrar Straus Giroux, 1962), pp. 337–386.

15. Mircea Eliade, *The Forge and the Crucible,* trans. Stephen Corrin, 2nd ed. (Chicago: University of Chicago Press, 1978), p. 151.

16. Sri Nisagadatta Maharaj, *I Am That* (Durham, N.C.: Acorn, 1997), p. 415.

17. The simplest account of Girard's ideas concerning mimesis is probably the interview with him at the end of René Girard, *To Double Business Bound: Essays on Literature, Mimesis, and Anthropology* (Baltimore: Johns Hopkins University Press, 1978). The fullest account is in René Girard, *Things Hidden since the Foundation of the World,* trans. Stephen Bann and Michael Metteer

(Stanford, Calif.: Stanford University Press, 1987), particularly Book 1, ch. 1, and Book 3, chs. 1 and 2. The most well-known account is in René Girard, *Violence and the Sacred,* trans. Patrick Gregory (Baltimore: Johns Hopkins University Press, 1977), in the chapter "From Mimetic Desire to the Monstrous Double."

18. Bertrand Russell, *Power* (London: Routledge, 2005), p. 23.

19. Aggression in lovemaking, the pleasure of making a partner have an orgasm or lose bodily or mental control, might function in a similar way, but it also points to power exchange as a kind of playing with the immanence of nonsensuous similarity.

20. Canetti, *Crowds and Power,* p. 382.

21. Eliade, *The Forge and the Crucible,* p. 91. Kieschnick makes note of the fact that those who made copies of sacred Buddhist texts in China also observed similar rituals of purification; see John Kieschnick, *The Impact of Buddhism on Chinese Material Culture* (Princeton: Princeton University Press, 2003), p. 174.

22. Eliade, *The Forge and the Crucible,* p. 141.

23. Canetti, *Crowds and Power,* p. 382. Irigaray, *Ethics of Sexual Difference,* p. 98.

24. "Original and Copycat," found on the Body Worlds website at www.bodyworlds.com/en/exhibitions/original_copycat.html (accessed March 20, 2010).

25. My thanks to Misha Yampolsky for pointing out to me the problematic of the shadow.

26. See Patrick Califia, "Sexual Politics, FTMs, and Dykes: Who Will Leap out of Bed First?" in Califia, *Speaking Sex to Power: The Politics of Queer Sex* (San Francisco: Cleis, 2002), pp. 107–120.

27. Judith Butler, *Giving an Account of Oneself* (New York: Fordham University Press, 2005). However, Butler's entire oeuvre, beginning with *Gender Trouble,* is a sustained meditation on transformative mimesis, via the theory of the performative discourses of identity and power.

4. Copying as Deception

1. Howard W. French, "Chinese Market Awash in Fake Potter Books," *New York Times,* August 1, 2007.

2. Ted Striphas, *The Late Age of Print: Everyday Book Culture, from Consumerism to Control* (New York: Columbia University Press, 2009), pp. 157–171.

3. Richard Posner, *The Little Book of Plagiarism* (New York: Pantheon, 2007), p. 19.

4. Www.harrypotterfanfiction.com, just one of a number of such sites, claims to hold more than 58,000 Harry Potter–related stories and to receive more than forty million hits per month.

5. Striphas, *The Late Age of Print.*

6. There is a vast and impressive scholarship on the topic of deception. Particularly useful to me were the global reviews in Mark Knapp, *Lying and Deception in Human Interaction* (Boston: Allyn and Bacon, 2008); Loyal Rue, *By the Grace of Guile: The Role of Deception in Natural History and Human Affairs* (New York: Oxford University Press, 1994); and *The Philosophy of Deception,* ed. Clancy Martin (Oxford: Oxford University Press, 2009). Also helpful was Don Herzog, *Cunning* (Princeton: Princeton University Press, 2006), which takes a pragmatic approach. Yet none of these books really carried me further than my own generalities/impasse in this paragraph.

7. Immanuel Kant, *Groundwork of the Metaphysic of Morals,* trans. H. J. Paton (New York: Harper and Row, 1964), pp. 70–71.

8. I follow, in part, the work of Henry Flynt in this abbreviated list. See, for example, "People Think" at www.henryflynt.org/human_relations/peoplethink.html (accessed March 20, 2010).

9. See Anthony Grafton, *Forgers and Critics: Creativity and Duplicity in Western Scholarship* (London: Collins and Brown, 1990), pp. 12, 13.

10. Sándor Radnóti, *The Fake: Forgery and Its Place in Art,* trans. Ervin Dunai (Lanham, Md.: Rowman and Littlefield, 1999), p. 51. Subsequent references to this work appear in parentheses in the text.

11. Patrick Radden Keefe, "The Jefferson Bottles: How Could One Collector Find So Much Rare Fine Wine?" *New Yorker,* September 3, 2007.

12. Slavoj Žižek, *Looking Awry* (Cambridge, Mass.: MIT Press, 2001), p. 74.

13. Roger Caillois, "Mimicry and Legendary Psychasthenia," in *The Edge of Surrealism: A Roger Caillois Reader,* trans. Claudine Frank and Camille Naish (Durham, N.C.: Duke University Press, 2003). Caillois later disavowed his comments as "fantastic" (*Man, Play, and Games,* trans. Meyer Barash [New York: Free Press of Glencoe, 1961], p. 178), but Lacan quotes them with

approval in *The Four Fundamental Concepts of Psycho-analysis,* trans. Alan Sheridan (London: Hogarth, 1977), pp. 73–74.

14. See Paul Virilio, *Desert Screen: War at the Speed of Light,* trans. Michael Degener (New York: Continuum, 2002). On Saddam's doppelgängers, see Tom Zeller, "Will the Real Saddam Hussein Please Step Down," *New York Times,* October 6, 2002.

15. Caillois, "Mimicry and Legendary Psychasthenia," p. 97.

16. Jacques Derrida, "History of the Lie: Prolegomena," in Derrida, *Without Alibi,* trans. Peggy Kamuf (Stanford: Stanford University Press, 2002), pp. 28–70.

17. Bertrand Russell, *Power* (London: Routledge, 2005), p. 23.

18. Walter Benjamin, "The Storyteller," in Benjamin, *Selected Writings,* vol. 3, ed. Howard Eiland and Michael W. Jennings (Cambridge, Mass.: Harvard University Press, 2002), p. 157.

19. Grafton, *Forgers and Critics,* p. 67.

20. Quoted in Robert E. Harrist Jr., "Replication and Deception in Calligraphy of the Six Dynasties Period," in *Chinese Aesthetics: The Ordering of Literature, the Arts, and the Universe in the Six Dynasties,* ed. Zongqi Cai (Honolulu: University of Hawaii Press, 2004), p. 31.

21. Www.buzzricksons.jp/detail.html (accessed August 22, 2007).

22. Glenn Gould, quoted in Christoph Cox and Daniel Warner, eds., *Audio Culture: Readings in Modern Music* (New York: Continuum, 2004), p. 121.

23. "Semblance and play form an aesthetic polarity. . . . This polarity must have a place in any definition of art. Art (the definition might run) is a suggested improvement on nature: an imitation that conceals within it a demonstration [of what the original should be]. In other words, art is a perfecting mimesis. In mimesis, tightly interfolded like cotyledons, slumber the two aspects of art: semblance and play." Walter Benjamin, "The Significance of Beautiful Semblance," in Benjamin, *Selected Writings,* vol. 3, p. 137.

24. Jeffrey Kripal, *Kali's Child: The Mystical and the Erotic in the Life and Teachings of Ramakrishna* (Chicago: University of Chicago Press, 1998).

25. See also the analysis of Philip K. Dick and *The Matrix* in Slavoj Žižek, *Welcome to the Desert of the Real* (London: Verso, 2002), ch. 1. Jean Baudrillard, *Simulations* (New York: Semiotext(e), 1983), is confusing because it vacillates between claiming that simulation is a strategy of deception employed by current political regimes (and thus analyzable in terms of history and truth) and the claim that there is, and always has been, only simulation as the

groundless constellation of the signs of the relative world—the position that Baudrillard later took up in books like *Fatal Strategies* and *Seduction.*

26. Jean Baudrillard, *Seduction,* trans. Brian Singer (New York: St. Martin's, 1990), pp. 180, 69–70. Subsequent references to this work appear in parentheses in the text.

27. François Jullien, *Treatise on Efficacy: Between Western and Chinese Thinking* (Honolulu: University of Hawaii Press, 2004), p. 192.

28. Philippe Lacoue-Labarthe, *Typography: Mimesis, Philosophy, Politics,* trans. Christopher Fynsk and others (Cambridge, Mass.: Harvard University Press, 1989), p. 297.

29. René Girard, *Evolution and Conversion: Dialogues on the Origins of Culture* (London: Continuum, 2007), pp. 211–214.

30. Two excellent guides to the mind-training practice are Chögyam Trungpa, *Training the Mind and Cultivating Loving-Kindness* (Boston: Shambhala, 1993); and Gomo Tulku, *Becoming a Child of the Buddhas: A Simple Clarification of the Root Verses of Seven Point Mind Training,* trans. Joan Nicell (Boston: Wisdom, 1998).

31. Lacan, *Four Fundamental Concepts of Psycho-analysis,* p. 6.

32. Morton Feldman, *Give My Regards to Eighth Street: Collected Writings of Morton Feldman* (Cambridge, Mass.: Exact Change, 2000), pp. 142–143.

5. Montage

1. Nicolas Bourriaud, *Relational Aesthetics,* trans. Simon Pleasance, Fronza Woods, and Mathieu Copeland (Dijon: Les Presses du Réel, 2002); and idem, *Postproduction,* trans. Jeanine Herman (New York: Lukas and Sternberg, 2002).

2. When Žižek calls Deleuze, the philosopher of the assemblage, "the ideologist of late capitalism," isn't he suggesting something similar? See Žižek, *Organs without Bodies: Deleuze and Consequences* (London: Routledge, 2004), p. 184.

3. Both Eisenstein and Vertov made their own lists of montage elements—lists that are striking because they almost exclusively concern the preparation for and planning of the montage; the actual making of the montage is bracketed in a single action. See Dziga Vertov, "From Kino-Eye to Radio-Eye," in *Kino-Eye: The Writings of Dziga Vertov,* trans. Kevin O'Brien (Berkeley: University of California Press, 1984), p. 90; Sergei Eisenstein, "The

Montage of Film Attractions," in Eisenstein, *Selected Writings*, vol. 1 (Bloomington: Indiana University Press, 1988), p. 50. Guy Debord and Gil Wolman have a list of the "laws of détournement" in their Situationist montage manifesto "A User's Guide to Détournement," though these "laws" mostly concern political principles of combination (*Situationist International Anthology*, ed. Ken Knabb [Berkeley: Bureau of Public Secrets, 2006], p. 18).

4. Douglas Kahn, *John Heartfield: Art and Mass Media* (New York: Tanam Press, 1985), p. 120.

5. Stanley J. Tambiah, "The Magical Power of Words," *Man*, n.s., 3, no. 2 (June 1968), pp. 175–208.

6. "Détournement as Negation and Prelude," in *Situationist International Anthology*, p. 67.

7. Eisenstein, "The Montage of Film Attractions," p. 50.

8. Kahn, *John Heartfield*, p. 112.

9. See Mikhail Yampolsky, "The Essential Bone Structure: Mimesis in Eisenstein," in *Eisenstein Rediscovered*, ed. Ian Christie and Richard Taylor (New York: Routledge, 1993), pp. 177–188. Also the chapter on Eisenstein and Vertov in P. Adams Sitney, *Modernist Montage: The Obscurity of Vision in Cinema and Literature* (New York: Columbia University Press, 1990).

10. Vertov, "From Kino-Eye to Radio-Eye," p. 90; Eisenstein, "The Montage of Film Attractions," p. 50.

11. Ruth McKendry, *Quilts and Other Bed Coverings in the Canadian Tradition* (Toronto: Van Nostrand Reinhold, 1979), pp. 99, 102.

12. Jean-Luc Nancy, quoted in Michel Gaillot, *Multiple Meaning: Techno, an Artistic and Political Laboratory of the Present* (Paris: Editions Dis Voir, 1998), p. 93.

13. Robert Farris Thompson, *Recycled, Reseen: Folk Art from the Global Scrap Heap* (New York: Abrams, 1996), p. 181.

14. Michael Taussig, *Defacement: Public Secrecy and the Labor of the Negative* (Stanford, Calif.: Stanford University Press, 1999), pp. 4, 25.

15. Henri Bergson, *Laughter: An Essay on the Meaning of the Comic*, trans. Cloudesley Brereton and Fred Rothwell (Rockville, Md.: Arc Manor, 2008; orig. pub. 1911), p. 21.

16. Michael Taussig, *The Magic of the State* (New York: Routledge, 1997), p. 3.

17. Simon Critchley, *Ethics, Politics, Subjectivity* (London: Verso, 1999), p. 235.

18. William S. Burroughs, *Nova Express* (New York: Grove), p. 53.

19. Alain Badiou, *The Century,* trans. Alberto Toscano (Cambridge: Polity, 2007), p. 48.

20. Timothy S. Murphy, "Exposing the Reality Film: William S. Burroughs among the Situationists," in *Retaking the Universe: William S. Burroughs in the Age of Globalization,* ed. Davis Schneiderman and Philip Walsh (London: Pluto, 2004), pp. 29–57; Guy Debord, *Society of the Spectacle* (Detroit: Black and Red, 1983), n.p., section 1, para. 34; Debord and Wolman, "A User's Guide to Détournement," p. 21.

21. Vertov, "From Kino-Eye to Radio-Eye," p. 88.

22. See Brion Gysin, "Cut-Ups: A Project for Disastrous Success," in *Back in No Time: The Brion Gysin Reader,* ed. Jason Weiss (Middletown, Conn.: Wesleyan University Press, 2001), pp. 125–132.

23. Standard accounts of collage that privilege modernism and all but ignore non-Western and premodern collage include Diane Waldman, *Collage, Assemblage, and the Found Object* (New York: Abrams, 1992); and Katherine Hoffman, ed., *Collage: Critical Views* (Ann Arbor, Mich.: UMI Research Press, 1989).

24. Lothar Ledderose, *Ten Thousand Things: Module and Mass Production in Chinese Art* (Princeton, N.J.: Princeton University Press, 2000).

25. John Storm Roberts, *Black Music of Two Worlds* (New York: Praeger, 1972), p. 11.

26. Roland Barthes, *Empire of Signs,* trans. Richard Howard (New York: Hill and Wang, 1982), p. 12.

27. "Mere listings of ingredients as in recipes, formulas, compounds, or prescriptions are not subject to copyright protection. However, when a recipe or formula is accompanied by substantial literary expression in the form of an explanation or directions, or when there is a combination of recipes, as in a cookbook, there may be a basis for copyright protection." See www.copyright.gov/fls/fl122.html (accessed July 10, 2008).

28. Dodie Bellamy, *Cunt-Ups* (New York: Tender Buttons, 2001), p. 65.

29. Eisenstein, "The Problem of the Materialist Approach to Form," in *Selected Writings,* vol. 1, p. 64.

30. Gayatri Spivak, *Death of a Discipline* (New York: Columbia University Press, 2003), p. 34.

31. Robin Lydenberg, "Engendering Collage: Collaboration and Desire in Dada and Surrealism," in Hoffman, *Collage: Critical Views,* pp. 271–285.

32. Luce Irigaray, *Je, tu, nous: Toward a Culture of Difference,* trans. Alison Martin (New York: Routledge, 1993), pp. 37–45.

33. Humberto Maturana and Francisco Varela, *Autopoiesis and Cognition: The Realization of the Living* (Boston: Reidel, 1980). Note also the work of Lynn Margulis and Dorion Sagan on the historical role of bacterial genes in the development of the human brain, *Microcosmos: Four Billion Years of Evolution from Our Microbial Ancestors* (Berkeley: University of California Press, 1997; orig. pub. 1986), p. 139ff.

34. Theodor Adorno, *Aesthetic Theory,* trans. Robert Hullot-Kentor (London: Continuum, 2004), p. 56.

35. Claude Lévi-Strauss, *The Savage Mind* (Chicago: University of Chicago Press, 1966), p. 17.

36. Jacques Derrida, "Structure, Sign, and Play in the Discourse of the Human Sciences," in Derrida, *Writing and Difference* (Chicago: University of Chicago Press, 1978), p. 285.

37. On postmodern and structuralist collage, see Thomas Brockelman, *The Frame and the Mirror: On Collage and the Postmodern* (Evanston, Ill.: Northwestern University Press, 2001); and Gregory L. Ulmer, "The Object of Post-Criticism," in Hal Foster, ed., *The Anti-Aesthetic: Essays on Postmodern Culture* (Port Townsend, Wash.: Bay Press, 1983), pp. 83–110.

38. Quoted in *The Wire,* 200 (October 2000), p. 36.

39. Norbert Wiener, *The Human Use of Human Beings: Cybernetics and Society* (Boston: Houghton Mifflin, 1954), pp. 3–4.

40. Brian Massumi, "On the Superiority of the Analog," in Massumi, *Parables for the Virtual: Movement, Affect, Sensation* (Durham, N.C.: Duke University Press, 2002), p. 135. Subsequent references to this work appear in parentheses in the text.

41. Jonathan Berger, interview with Nora Young on CBC radio, www.cbc.ca/spark/2009/03/full-interview-jonathan-berger-on-mp3s-and-sizzle (accessed October 24, 2009).

6. The Mass Production of Copies

1. See Lori Pauli, *Manufactured Landscapes: The Photographs of Edward Burtynsky* (Ottawa: National Gallery of Canada in association with Yale University Press, 2003), pp. 101–115.

2. Michael Taussig, *Mimesis and Alterity: A Particular History of the Senses*

(New York: Routledge, 1993), pp. 59–70; Max Horkheimer and Theodor W. Adorno, *Dialectic of Enlightenment,* trans. John Cumming (New York: Herder and Herder, 1972), p. 185.

3. Lothar Ledderose, *Ten Thousand Things: Module and Mass Production in Chinese Art* (Princeton, N.J.: Princeton University Press, 2000), pp. 78–79. *Brill's New Pauly: Encyclopaedia of the Ancient World,* ed. Hubert Cancik and Helmut Schneider (Boston: Brill, 2002–), vol. 1, entry for "amphora," pp. 614–615.

4. Karl Marx, in *The Marx-Engels Reader,* ed. Robert C. Tucker, 2nd ed. (New York: Norton, 1978), p. 320.

5. Taussig, *Mimesis and Alterity,* pp. 22–23.

6. Arjun Appadurai, "Introduction: Commodities and the Politics of Value," in Appadurai, ed., *The Social Life of Things: Commodities in Cultural Perspective* (Cambridge: Cambridge University Press, 1986), pp. 3–63.

7. Glyn Davies, *A History of Money* (Cardiff: University of Wales Press, 2002), pp. 34–65; Jonathan Williams, ed., *Money: A History* (London: British Museum Press, 1997), p. 23.

8. Naomi Klein, *No Logo: Taking Aim at the Brand Bullies* (Toronto: Knopf Canada, 2000).

9. Benoît Heilbrunn distinguishes between "branding" and "badging," the former meaning the building up of the identity of a particular company, based on the products that the public associates them with, and the latter meaning the conferring of the brand's prestige on a variety of other products that have little to do with what the company was originally known for (for example, Louis Vuitton sunglasses). Benoît Heilbrunn, "Le luxe est mort, vive le luxe! Le marché du luxe à l'aune de la democratisation," in *Le luxe: Essais sur la fabrique de l'ostentation,* ed. Olivier Assouly (Paris: Editions du Regard, 2005), pp. 361–364.

10. Jean Baudrillard, *The System of Objects,* trans. James Benedict (New York: Verso, 2005), p. 100.

11. Rokeby's key paper on the topic is "The Construction of Experience: Interface as Content," homepage.mac.com/davidrokeby/experience.html (accessed October 30, 2008).

12. Morton Feldman, *Give My Regards to Eighth Street: Collected Writings of Morton Feldman* (Cambridge, Mass.: Exact Change, 2000), pp. 148–149.

13. Cited in Dalia Judovitz, *Unpacking Duchamp: Art in Transit* (Berkeley: University of California Press, 1995), p. 124.

14. Ibid., p. 129. The quote from Duchamp can be found in Marcel Duchamp, *Notes* (Boston: Hall, 1983), n.p., note 18. Consider also the following quote (ibid., note 35):

> 2 forms cast in / the same mold(?) differ / from each other / by an infra thin separative / amount—
>
> All "identicals" as / identical as they may be, (and / the more identical they are) / move toward this / infra thin separative / difference.
>
> Two men are not / an example of identicality / and to the contrary move away / from a determinable / infra thin difference—but . . .
>
> [verso] there exists the crude conception / of the déjà vu which leads from / generic grouping / (2 trees, 2 boats) / to the most identical "castings" / It would be better / to try / to go / into the / infra thin / interval which separates / 2 "identicals" than / to conveniently accept / the verbal generalization / which makes / 2 twins look like 2 / drops of water.

15. Ibid., note 1.

16. Benjamin's best-known discussion of aura is contained in sections III and IV of "The Work of Art in the Age of Its Technological Reproducibility," in Walter Benjamin, *Selected Writings,* vol. 3, ed. Howard Eiland and Michael W. Jennings (Cambridge, Mass.: Harvard University Press, 2002), pp. 103–105. But there are many other references in Benjamin's work, as documented in Miriam Bratu Hansen, "Benjamin's Aura," *Critical Inquiry,* 34, no. 2 (Winter 2008), pp. 336–375.

17. "In Time the same object is not the / same after a 1 second interval—what / Relations with the identity principle?" Duchamp, *Notes,* note 7.

18. Marcel Duchamp, "Apropos of 'Readymades,'" in David Evans, ed., *Appropriation* (Cambridge, Mass.: MIT Press, 2009), p. 40.

19. Bruce Sterling, *Shaping Things* (Cambridge, Mass.: MIT Press, 2005), p. 79. Subsequent references to this work appear in parentheses in the text.

20. Paper consumption has increased since the introduction of computers; see www.cbc.ca/technology/story/2006/11/10/tech-paperless.html?ref=rss (accessed March 5, 2010).

21. Norbert Wiener, *The Human Use of Human Beings: Cybernetics and Society* (Boston: Houghton Mifflin, 1954), pp. 109–110.

22. Julian Dibbell, personal conversation, November 29, 2005.

23. Henry Flynt, "Spirituality: A Critical Account," unpublished, 2007.

24. My account here draws on Lisa Trivedi, *Clothing Gandhi's Nation: Homespun and Modern India* (Bloomington: Indiana University Press, 2007). My thanks to Sunita Kumar for pointing out the *charka*'s politics of mimesis.

25. Steve Goodman, *Sonic Warfare: Sound, Affect, and the Ecology of Fear* (Cambridge, Mass.: MIT Press, 2010), p. 171. Subsequent references to this work appear in parentheses in the text.

7. Copying as Appropriation

1. Benjamin H.D. Buchloh, "Parody and Appropriation in Francis Picabia, Pop and Sigmar Polke," reprinted in *Appropriation,* ed. David Evans (Cambridge: MIT Press, 2009), p. 180.

2. In general, appropriation remains a severely underresearched topic. An important exception is Rosemary Coombe, *The Cultural Life of Intellectual Properties* (Durham, N.C.: Duke University Press, 1998).

3. Zhang Boduan, "Essay on Achieving Perfection," cited in Zhang Jiyu and Li Yuanguo, "'Mutual Stealing among the Three Powers' in the Scripture of Unconscious Unification," in *Daoism and Ecology: Ways within a Cosmic Landscape,* ed. N. J. Girardot, James Miller, and Liu Xiaogan (Cambridge, Mass.: Harvard University Press, 2001), p. 121.

4. Lynn Margulis and Dorion Sagan, *Microcosmos: Four Billion Years of Evolution from Our Microbial Ancestors* (Berkeley: University of California Press, 1997), p. 33.

5. Yang Jwing-Ming, *Advanced Yang Style Tai Chi Chuan* (Jamaica Plain, N.Y.: Yang's Martial Arts Academy, 1987), pp. 4–5.

6. See Judith Simmer-Brown, *Dakini's Warm Breath: The Feminine Principle in Tibetan Buddhism* (Boston: Shambhala, 2001), pp. 121–127.

7. My understanding of historical theories of property and their relationship to appropriation has been stimulated by my reading of Martin Zeilinger, "Art and Politics of Appropriation" (diss., University of Toronto, 2009) and Zeilinger's discussion of Peter Drahos, *A Philosophy of Intellectual Property* (Aldershot, U.K.: Dartmouth, 1996).

8. John Locke, *Second Treatise on Government* (Indianapolis: Hackett, 1980), p. 20.

9. The passage is from "Economic and Philosophical Manuscripts of 1844," in Marx's *Early Writings* (London: Penguin, 1992), p. 351. I quote Allen Wood's translation in *Karl Marx* (London: Routledge, 2004), p. 40.

10. Caleb Crain, "Bootylicious," *New Yorker,* September 7, 2009, p. 74.

11. Bernard Edelman, *Ownership of the Image: Elements for a Marxist Theory of Law,* trans. Elizabeth Kingdom (London: Routledge and Kegan Paul, 1979), p. 46. Subsequent references to this work appear in parentheses in the text.

12. Timothy Morton, *Ecology without Nature: Rethinking Environmental Aesthetics* (Cambridge, Mass.: Harvard University Press, 2007).

13. Homi Bhabha, *The Location of Culture* (New York: Routledge, 2004), pp. 121–131.

14. See, for example, James Young and Conrad Brunk, eds., *The Ethics of Cultural Appropriation* (Oxford: Wiley-Blackwell, 2009).

15. Coco Fusco, "Who's Doin' the Twist? Notes toward a Politics of Appropriation," in *English Is Broken Here: Notes on Cultural Fusion in the Americas* (New York: New Press, 1995), pp. 65–77.

16. Eric Gable, "An Anthropologist's (New?) Dress Code: Some Brief Comments on a Comparative Cosmopolitanism," *Cultural Anthropology,* 17, no. 4 (November 2002), pp. 572–579.

17. *Bamboozled* script at www.imsdb.com/scripts/Bamboozled.html (accessed August 4, 2008).

18. Gayatri Chakravorty Spivak, "Righting Wrongs," *South Atlantic Quarterly,* 103, nos. 2–3 (Spring–Summer 2004), p. 532.

19. In *Anti-Oedipus: Capitalism and Schizophrenia* (New York: Viking, 1977), Gilles Deleuze and Félix Guattari argue, for example, that traditional societies were structured precisely to avoid or refuse the impasse of capital.

20. Martin Heidegger, "The Way to Language," trans. Peter D. Hertz, in Heidegger, *On the Way to Language* (New York: Harper and Row, 1971), p. 129, footnote. Stambaugh's note on *Ereignis* can be found in the preface to Martin Heidegger, *Identity and Difference* (Chicago: University of Chicago Press, 2002), p. 14. Richard Polt reaches a similar conclusion concerning the correct translation of *Ereignis* in his entry on the term in *A Companion to Heidegger,* ed. Hubert Dreyfus and Mark Wrathall (Oxford: Blackwell, 2005);

and in Richard Polt, *The Emergency of Being: On Heidegger's Contributions to Philosophy* (Ithaca: Cornell University Press, 2006).

21. See the discussion in Graham Harman, "Dwelling with the Four-Fold," *Space and Culture,* 12, no. 3 (2009), pp. 292–302, although I disagree with the project of transforming Heideggerian philosophy into one that is entirely object-centered.

22. See Martin Heidegger, "The Thing" and ". . . Poetically Man Dwells . . . ," in Heidegger, *Poetry, Language, Thought,* trans. Albert Hofstadter (New York: HarperCollins, 2001), pp. 161–184 and 209–227.

23. Karl Marx, *Grundrisse: Foundations of the Critique of Political Economy,* trans. Martin Nicolaus (London: Penguin, 1993), pp. 87–88.

24. Friedrich Nietzsche, *Beyond Good and Evil,* trans. R. J. Hollingdale (London: Penguin Books, 2003), p. 194.

25. Georges Bataille, "The Use Value of the Marquis de Sade," in Bataille, *Visions of Excess: Selected Writings, 1927–1939,* trans. Allan Stoekl, with Carl R. Lovitts and Donald M. Leslie Jr. (Minneapolis: University of Minnesota Press, 1985), pp. 91–102.

26. Hélène Cixous, "The Laugh of the Medusa," in *The Norton Anthology of Theory and Criticism,* ed. Vincent B. Leitch (New York: Norton, 2001), p. 2052.

27. Simone Weil, "Decreation," in *The Simone Weil Reader,* ed. George A. Manichas (Mt. Kisco, N.Y.: Moyer Bell, 1985), pp. 350–356.

28. Lucy R. Lippard, ed., *Six Years: The Dematerialization of the Art Object from 1966 to 1972* (Berkeley: University of California Press, 1997). See also Peggy Phelan, *Unmarked: The Politics of Performance* (London: Routledge, 1993).

29. Hakim Bey, *T.A.Z.: The Temporary Autonomous Zone, Ontological Anarchy, Poetic Terrorism* (Brooklyn, N.Y.: Autonomedia, 1991).

30. Simon Reynolds, *Generation Ecstasy: Into the World of Techno and Rave Culture* (London: Routledge, 1999), p. 291. Reynolds also uses the term "cheesiness" in various articles.

31. Fredric Jameson, "Fear and Loathing in Globalization," *New Left Review,* 23 (September 2003).

32. McKenzie Wark, *A Hacker Manifesto* (Cambridge, Mass.: Harvard University Press, 2004).

33. Jacques Derrida, *Given Time, I: Counterfeit Money* (Chicago: University

of Chicago Press, 1992). Derrida's book explicitly treats copying at several points, since his topic is in part "counterfeit money" and the problem of the fake. For example: "In 'Counterfeit Money,' on the other hand, it is a matter of (perhaps legitimate, one will never know) children or (perhaps real and good) interest produced not from an Idea, or even from the Idea of the Good, from true Capital, or from the true Father, not even from a copy of the idea, from an icon or idol, for example a (monetary, conventional, and artificial) sign, but from a simulacrum, from a copy of a copy (phantasma). The phantasm is recognized as having the power, at least the power and the possibility—without any controlling certitude, without any possible assurance—of producing, engendering, giving" (157). The following discussion of engendering and production, while not directly focused on copying, is highly relevant to the topic.

34. Moore, *Mix Tape,* p. 13.

35. John Giorno, "Everyone Gets Lighter," in Giorno, *Subduing Demons in America: Selected Poems, 1962–2007* (Berkeley, Calif.: Soft Skull / Counterpoint, 2008), pp. 352–353.

36. Www.dalailama.com/news.42.htm (accessed October 23, 2009).

Coda

1. R. C. Barrington Partridge, *History of the Legal Deposit of Books throughout the British Empire* (London: Library Association, 1938).

2. See, for example, Kenyon Wallace, "Textbook Piracy Thriving around City's Campuses," *Toronto Star,* January 10, 2009, www.thestar.com/News/GTA/article/568628 (accessed February 21, 2010); Elizabeth Kagedan and Tim Legault, "Busted!" *Varsity,* October 19, 2009, thevarsity.ca/articles/21369 (accessed February 21, 2010); and Access Copyright's press release for the Quality Control Copy bust, www.accesscopyright.ca/docs/News%20Releases/15Oct2009.pdf (accessed February 21, 2010).

3. See www.accesscopyright.ca/Default.aspx?id=35 (accessed February 21, 2010).

4. On fair dealing in Canada, see Samuel Trosow and Laura Murray, *Canadian Copyright: A Citizen's Guide* (Toronto: Between the Lines, 2007), pp. 74–89. It is worth noting that the word "deal" also has an interesting history, originating in medieval times to describe a process of sharing and distribution, and taking on a specifically trade-related meaning only in the Renaissance.

5. Kate Eichhorn, "Breach of Copy/rights: The University Copy District as Abject Zone," *Public Culture,* 18, no. 3 (2006), p. 555.

6. See Martin Zeilinger, "Art and Politics of Appropriation" (diss., University of Toronto, 2009), for a review of the uses of appropriation. See also John Locke, *Second Treatise on Government* (Indianapolis: Hackett, 1980), p. 20; Karl Marx, "Economic and Philosophical Manuscripts of 1844," in Marx, *Early Writings* (London: Penguin, 1992), p. 351.

7. See Rosemary Coombe, *The Cultural Life of Intellectual Properties: Authorship, Appropriation, and the Law* (Durham, N.C.: Duke University Press, 1998); and McKenzie Wark, *A Hacker Manifesto* (Cambridge, Mass.: Harvard University Press, 2004).

8. Georges Bataille, *The Accursed Share,* vol. 1, trans. Robert Hurley (New York: Zone, 1988).

9. En.wikipedia.org/wiki/Captain_Copyright (accessed February 22, 2010).

10. *Hegel's Philosophy of Right,* trans. T. M. Knox (Oxford: Oxford University Press, 1967), pp. 37ff.

11. Hakim Bey, *T.A.Z.: The Temporary Autonomous Zone, Ontological Anarchy, Poetic Terrorism* (Brooklyn, N.Y.: Autonomedia, 1991).

12. Michael Hardt and Antonio Negri, *Empire* (Cambridge, Mass.: Harvard University Press, 2000); idem, *Multitude* (New York: Penguin, 2004); and idem, *Commonwealth* (Cambridge, Mass.: Harvard University Press, 2009).

Acknowledgments

As befits a book written "in praise of copying," I have many people to thank for what appears here under my name. For ideas, conversation, support, comments on various drafts, inspiration, creative opposition, and much more, my thanks go to: Spiros Antonopoulos, Martin Arnold, Barbara Balfour, Ian Balfour, David Banash, Cary Berger, Ritu Birla, Alan Bishop, the Boon family (especially Derek), Eric Cazdyn, Eric Chenaux, Jace Clayton, Rosemary Coombe, Erik Davis, Julian Dibbell, Cheryl Donegan, Dorian and Dorian (a.k.a. Alex Livingston and Merike André-Barrett), Henry Flynt, FM3, Jay Garfield, John Giorno, Kenneth Goldsmith, David Goldstein, Richard Grant, Ruth Hawkins, Tim Hecker, Louis Kaplan, Sri Karunamayee, Michele Laporte, Alan Licht, Clea McDougall and all at the late lamented *Ascent*, Kembrew McLeod, Raz Mesinai, Tim Morton, Richard Nash, Nicholas Noyes, Aki Onda, the Pearson and Patch families, Sarah Peebles, Liz Phillips, Dawson Prater, Anais Prosaic, Ben Ratliff, Trace Reddell, Avital Ronell, Marina Rosenfeld, Joseph Sonnabend, Sparrow, Chrysanne Stathacos, Louis Suarez-Potts, Michael Taussig, Andrew Wedman, Tobias van Ween, Darren Wershler, John Whalen-Bridge, Jennifer Wicke, Peter Lamborn Wilson, everyone at *The Wire*, Misha Yampolsky, La Monte Young, Marian Zazeela, and Martin Zeilinger.

Thanks to my Buddhist teachers Gehlek Rimpoche, H.H., the Dalai Lama, Khenpo Tsültrim Gyatso, and Khenpo Sonam Tobgyal for sharing their knowledge of the Buddhadharma with me. I hope that some traces of their profound teachings have found their way into this book.

Thanks to the students in my undergraduate and graduate classes on sampling and contemporary literature at York University, with whom I discussed many of the ideas and problems addressed in this book. In particular, for their comments and questions: Liza Chiber, Tyler Crick, Tom Cull, Francesca D'Angelo, Jason Demers, Erin Gray, Sabina Kim, Sean Kolenko, Sunita Kumar, Yves Lamson, Martin Leduc, Rea McNamara, Mark Pascual, Jacqueline St. Bernard, Terry Sudeyko, Sarah Trimble, Stephen Voyce, Suzanne Zelazo.

Thanks also to my agent Paul Bresnick, who first proposed that I write a book about contemporary cultures of copying; to Kenneth Goldsmith, with whom I wrote a proposal for said book that mutated into our separate projects; to my editor Lindsay Waters for patience, support, and trust in my ability to bring this off; to Phoebe Kosman, Hannah Wong, and Maria Ascher at Harvard for their careful work on behalf of the book; to my graduate assistants Tom Cull and Stephen Voyce for their diligent research and fact checking on my behalf; to Kim Michasiw, Julia Creet, and Art Redding for their support of my work when they were departmental chairs at York; and to the anonymous peer reviewers at Harvard University Press for challenging me to fully develop my arguments.

I gratefully acknowledge support of an SSHRC Standard Research Grant from the Canadian government, and a York University Faculty of Arts Fellowship for the study of Asian religions and literature, a topic which improbably assumed central importance in this book, as well as York Faculty of Arts Research Grants and Specific Research Grants, which also provided assistance as I completed various aspects of research on this project.

Finally, thanks to my wife Christie Pearson and my son Jesse for their patience, love, and support through the many iterations, copies, and transformations that the manuscript passed through. This book is the product of their sacrifice and joy, and one small manifestation of my love for them.

Index